# DAVIS MOUNTAINS VISTAS

# DAVIS MOUNTAINS VISTAS

A Geological Exploration of the Davis Mountains

By

William MacLeod

TEXAS GEOLOGICAL PRESS
ALPINE, TEXAS

Copyright © 2005 by William MacLeod
All rights reserved.

First Printing, December 2005

Printed and Bound by Capital Printing Company
Austin, Texas

Cover Photo: "Sawtooth Mountain" by William MacLeod

Cover Design: Leisha Israel, digital tractor
Bastrop, Texas

**Publishers Cataloging in Publication Data**
MacLeod, William
    Davis Mountains Vistas: a geological exploration
    of the Davis Mountains / by William MacLeod
        p.   cm.
    Includes bibliographic references and index
    ISBN 0-9727786-0-0
    1. Geology – Davis Mountains Region (Tex.).   I. Title.
QE168.B5 M24 2005
557.64-dc22

TEXAS GEOLOGICAL PRESS
P.O. Box 967
Alpine, Texas 79831
www.texasgeologicalpress.com

## Contents

**Chapter**

| | | |
|---|---|---|
| | Introduction | 7 |
| 1 | Using the Book | 11 |
| 2 | Development of the Davis Mountains | 12 |
| 3 | Approaching the Mountains | 36 |
| 4 | Toyahvale – Fort Davis | 48 |
| 5 | Fort Davis – Kent | 67 |
| 6 | Kent – Toyahvale | 96 |
| 7 | The Davis Mountains Scenic Loop | 100 |
| 8 | Fort Davis – Alpine – Marfa –Fort Davis | 121 |
| 9 | The Salt Basin Rift | 148 |
| 10 | Kent – Van Horn – Valentine – Marfa | 159 |
| | Glossary | 178 |
| | Reading List | 183 |
| | Index | 187 |

*For Martha*

# Introduction

The Davis Mountains evoke cool, airy expanses of mountainous terrain much appreciated by Texans in summertime. They are the first of the southwestern mountain ranges you see when traveling west across Texas, standing on the edge of the great North American cordillera, but not part of it, on the boundary between the Great Plains and the Rocky Mountains. They are part of the Trans-Pecos volcanic field, the easternmost of the southwest volcanic fields. The field runs from Interstate 10 north of Fort Davis down to the Rio Grande and across into Mexico and includes the volcanic rocks of the Chisos and Chinati Mountains.

This book attempts to tell the story of the Davis Mountains for the general reader. It describes their geology in terms understandable by visitors who know little about the subject, but in order to avoid tedious repetitions of definitions, a glossary is included in which most technical terms are defined.

The mountains are built up of layer upon layer of volcanic rocks erupted from fissures and vents, some of which have been identified, others of which are buried under younger rocks. When volcanoes are mentioned, visions of smoking pyramids come to mind but the volcanoes of the American southwest were not of that kind. Rather than having central vents, they consisted of fissures and vents through which molten rock oozed or exploded to the surface. In the Davis Mountains, five main lava episodes alternated with explosive ash episodes, the latter mainly resulting from water coming into contact with the molten rock.

In the Paisano volcano between Alpine and Marfa, which produced three main lava layers with interbedded ash layers, much of the upper lava layer has been eroded away, laying bare the pattern of eruption fissures, now filled with solidified magma. There may be as many as a thousand such fissures in this volcano.

The final lavas in the Davis Mountains were viscous products that formed volcanic domes on the summits of Mount Livermore and Brooks Mountain, the two highest peaks in the range. Both have been eroded, the Mount Livermore one into a craggy ridge culminating in the high bare Baldy Peak. The lavas that created the volcanic domes are most likely the surface manifestations of a large intrusion that uplifted older volcanic layers around them as much as 2,000 feet.

Once the main volcanic phase ended, the area to the southwest of the mountains underwent a period of stretching in which a series of basins developed along a trend from New Mexico to around Marfa. The basins define a young or not very well developed rift valley which in this book is called the Salt Basin Rift after its deepest basin. Much of the main Davis Mountains was tilted into the rift.

Strictly speaking, the Davis Mountains are those mountains west and north of Fort Davis but the book includes descriptions of the mountains around Alpine and the Barrilla Mountains as their geology is inseparable from that of the Davis Mountains. The final two chapters describe the Salt Basin Rift, a subject of outstanding scientific interest.

Although the volcanic field has been studied by several generations of geologists, dating from at least 1889, when Professor von Streeruwitz traveled through the area by train collecting specimens on behalf of the Texas Geological Survey, much remains to be done. Volcanic rocks are quite difficult to correlate between occurrences. There are no easy diagnostic tools, such as fossils provide in sedimentary rocks, to distinguish one volcanic rock from another. Most of the units in the Davis Mountains are of very similar chemical composition and closely resemble each other. Adding to the difficulty, volcanic rocks can change character from one point to another. It will take a concerted effort using perhaps trace element analysis and isotopic dating to develop reliable correlations.

Another problematic scientific issue in the Davis Mountains is the difficulty of distinguishing between lavas and rocks derived from ash flows. Several of the lava units cover enormous areas compared to silica-rich lavas in general and many geologists suspect them of being ash-flow units in which evidence of their origins has been obliterated.

Turning to specific conventions used in this book, researchers have named the more extensive units among the large number of flows in the mountains, resulting in a great catalog of formation names. These are difficult to remember, so as well as creating a table of the entire set of formations in Chapter 2, I have provided a short table at the beginning of each chapter detailing the formations you will see in that particular chapter.

# INTRODUCTION

In naming formations I have followed the conventions of the Bureau of Economic Geology, except when formations have been revised. For example, Henry, Kunk and McIntosh have revised the definitions of the Barrel Springs-Wild Cherry formations and have recognized an informal unit which they call the tuffs of Wild Cherry. I have used the term Wild Cherry tuffs for this unit. In referring to mountains or mountain ranges, I include the height of the summit in brackets after the first reference to the name, as in "Sawtooth Mountain (7,686 feet)", or in the case of a range, the highest point, as in "Apache Mountains (5,216 feet)".

When describing a feature or in the caption to a diagram, I often include names of authors who have researched the topic. In such cases, the author or authors and a citation for the source article or book can be found in the reading list at the end of the book, indexed by author. The preeminent author in the list is Don Parker of Baylor University, and all of us who are interested in the Davis Mountains owe a debt of gratitude to him for his work and the work of his students over the past 30 years. Christopher Henry and Jonathon Price, formerly with the Bureau of Economic Geology, have also contributed greatly to our understanding of the area.

I would like to thank all those who contributed to the development of this book, especially Blaine Hall of Sul Ross State University, who has been enormously helpful as a sounding board for my theories as they have developed. His private collection of geological publications and his references from the University interlibrary loan program have been very beneficial as was his critical reading of much of the manuscript.

I would also like to thank Pat Dasch, who was kind enough to edit the manuscript, and geologists Julius Dasch and Scott Baldridge who read and greatly improved it. Also, thanks to John Karges of The Nature Conservancy, Kathryn Hoyt of CDRI, Mark Cash of McDonald Observatory, David Bischhausen of Davis Mountains State Park, Tom Johnson of Balmorhea State Park and Don Mulhern, all of whom read parts of the manuscript. Finally, I thank my wife Martha, to whom this book is dedicated, for her help and support as editor-in-chief throughout its development. She holds a steady course in all seasons.

*William MacLeod*
*Alpine, Texas*
*November 29, 2005*

1: USING THIS BOOK

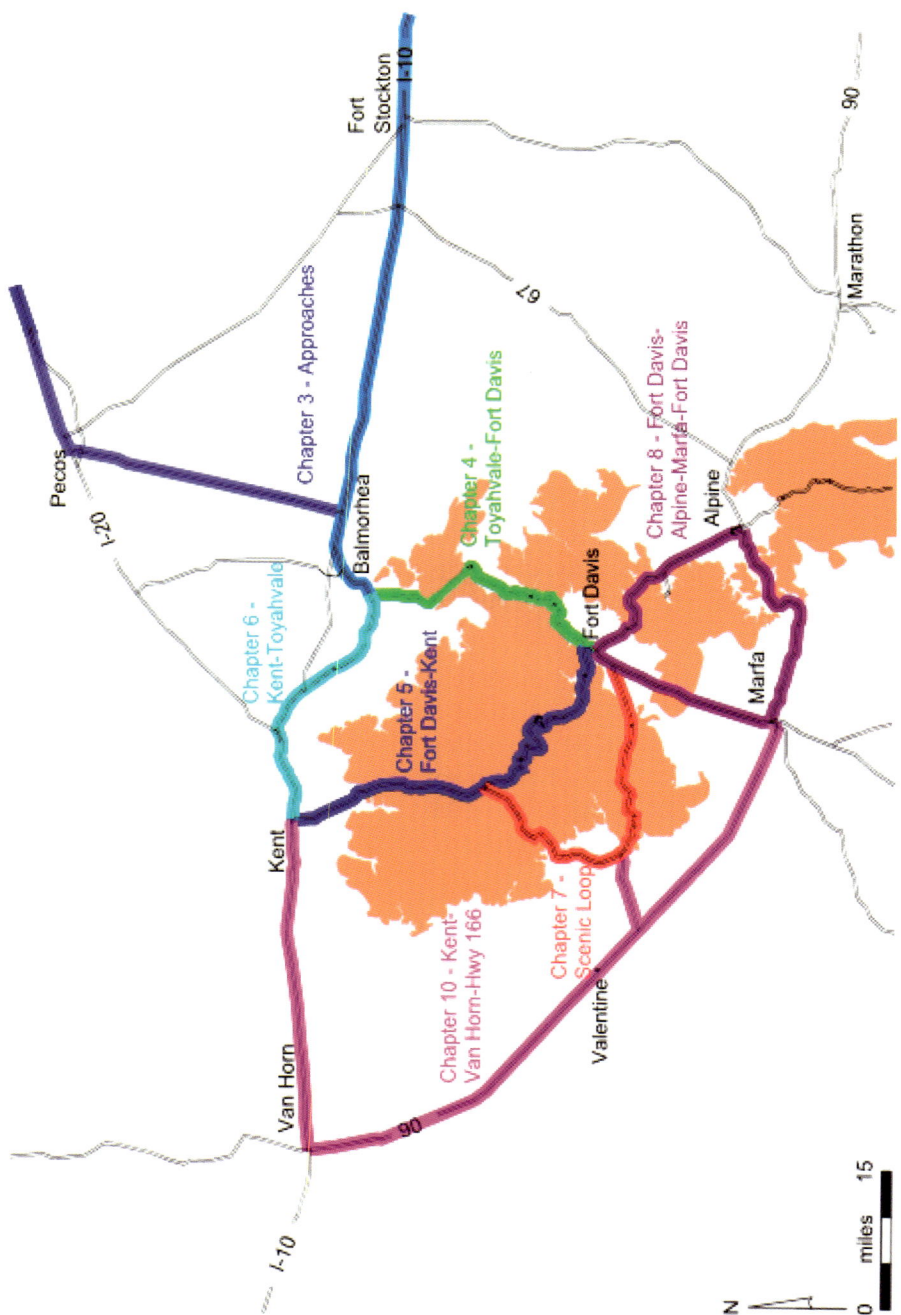

Fig. 1.1: Pick your own routes from the road segments described in Chapters 3-8 and 10.

# 1: Using this Book

The book is divided into two sections, the first describing the Davis Mountains, the second the Salt Basin Rift, a rift valley to the west of the mountains, little written about, but of great scientific and scenic interest. Each section has an introductory chapter giving the geological history followed by chapters describing road trips around the area. The four main loops you can take through and around the mountains are:

**Fort Davis – Kent – Toyahvale – Fort Davis (112 miles):** This is my personal favorite. It takes you through the heart of the mountains past the Davis Mountains State Park and the McDonald Observatory, both well worth visiting, then along Interstate-10 north of the mountains and back up to Fort Davis along Limpia Canyon. Restaurant food is available at the McDonald Observatory cafeteria or Balmorhea. Convenience supplies and gasoline are available at Kent. Allow at least 4 hours for this trip.

**The Scenic Loop (75 miles):** This route follows the Kent road for 29 miles, then swings left round the southwest flank of the mountains, taking about 3 hours. Food is available at the McDonald Observatory cafeteria.

**Fort Davis – Alpine – Marfa – Fort Davis (71 miles):** This loop takes you down Musquiz Canyon to Alpine, then through the Paisano volcano to Marfa, taking about 3 hours. Food and shopping are available in Alpine and Marfa.

**Fort Davis – Kent – Van Horn – Marfa – Fort Davis (130 miles):** This route repeats the journey from Fort Davis to Kent, and then travels along Interstate 10 to Van Horn between the Apache and Wylie Mountains. The leg from Van Horn to Marfa is down the Salt Basin Rift between the Van Horn Mountains-Sierra Vieja and the Wylie and Davis Mountains. The entire loop takes about 4 to 6 hours. Food is available at the McDonald Observatory, Van Horn and Marfa.

# 2: Development of the Davis Mountains

The Davis Mountains volcanic field forms a northwest-southeast belt, 35 miles wide and 80 miles long stretching from Interstate 10 near Kent to Elephant Mountain, 30 miles south of Alpine (Fig. 2.1), where it merges with volcanic rocks from the Chinati Mountains volcanic center. The field was created over a 3-million year period beginning around 38 million years ago.

The mountains are built up of layers of volcanic lava and ash and scattered igneous intrusions. In Fig. 2.2, the layers as they appear today are shown superimposed on one another. The largest number of layers occurs around the highest mountains, as you would expect, and they tail off to the south and north.

Layers range in thickness from as much as 1,000 feet for lava units to as little as a tenth of an inch in ash units. Source fissures and vents for some units have been identified but most are buried under younger layers. In places, calderas or volcanic depressions developed where underground reservoirs emptied and their roofs collapsed. Once volcanic activity ended, erosion took over, stripping the mountains of their upper layers and carving deep canyons in from their peripheries.

The mountains are bordered on their southwest by a broad plain from Van Horn to Marfa that separates them from the Sierra Vieja and Van Horn Mountains. The plain overlies a rift valley, a zone in which blocks of strata have dropped down into basins, some of them 3,000 feet deep. The basins are now buried by lake and stream deposits of sands and gravels eroded from the mountains.

The remainder of this chapter discusses geological basics, the classification of rocks, the mechanisms of plate tectonics, and geological time. It then goes on to describe the history and structure of the Davis Mountains and their surroundings up to the end of the major phase of volcanism. Later events, the development of the rift valley and erosion of the mountains, are discussed in Chapter 9.

## Classification of Rocks

In geology, a rock is any natural material, soft or hard, consisting of one or more minerals. Rocks are broadly classified into three categories, igneous, sedimentary, and metamorphic.

The majority of the rocks in the Davis Mountains are igneous, i.e. they solidified from molten or partly molten material called magma that forms underground above hotter regions in the Earth's interior. Magma may contain suspended solids such as crystals and rock fragments. If it erupts on to the Earth's surface, it forms *volcanic rocks*. Magma solidifying below the Earth's surface creates *igneous intrusions*.

A magma's chemical composition determines its behavior when extruded. Silica-rich or *felsic* magmas are more viscous than iron- and magnesium-rich *mafic* magmas. Most Davis Mountains magmas were felsic creating rhyolite and trachyte (Fig. 2.10). Felsic magma erupted as molten lava is usually quite viscous, rather like oatmeal; it melts at about 750 degrees Centigrade. If it contains water, the superheated water will cause the magma to explode on release of pressure when it reaches the surface, creating great plumes of dust and ash. Felsic magmas usually become drier as eruptive activity continues, and later lavas ooze out as thick flows and domes. Rocks derived from explosive eruption are called *pyroclastics*.

Volcanic ash is made up of particles of glass or pumice. Glass is unstable and crystallizes over time into mainly clay minerals creating a mixture of glass and clay called *tuff*. Several kinds of tuff are found in the Davis Mountains. One common kind is *ash-fall tuff*, which, as the name implies, is derived from airborne volcanic ash falling to earth. Another variety is *ash-flow tuff*. An ash flow is a density current, a combination of hot air and ash that travels along the ground from a volcanic fissure or vent. It can travel great distances down slopes and often up quite steep slopes. Eventually, the flow loses momentum and builds up a layered ash-flow deposit that turns into tuff over time.

Ash deposits and tuffs are easily eroded by rain or flowing water which redeposits them as sandstone or mudstone; such sedimentary rocks made up of volcanic rock fragments are called *volcaniclastic rocks*. Sometimes an ash flow is so hot that when it settles it welds together by the combined action of the heat retained by particles, the weight of overlying material, and hot gases. This *welded tuff* can be as hard and resistant to erosion as lava, and is often difficult to distinguish from lava in outcrop.

Several of the Davis Mountains lavas appear to have spread 20 or 30 miles from their sources although, worldwide, most felsic lavas travel only a few miles from their sources.

## 2: DEVELOPMENT OF THE MOUNTAINS

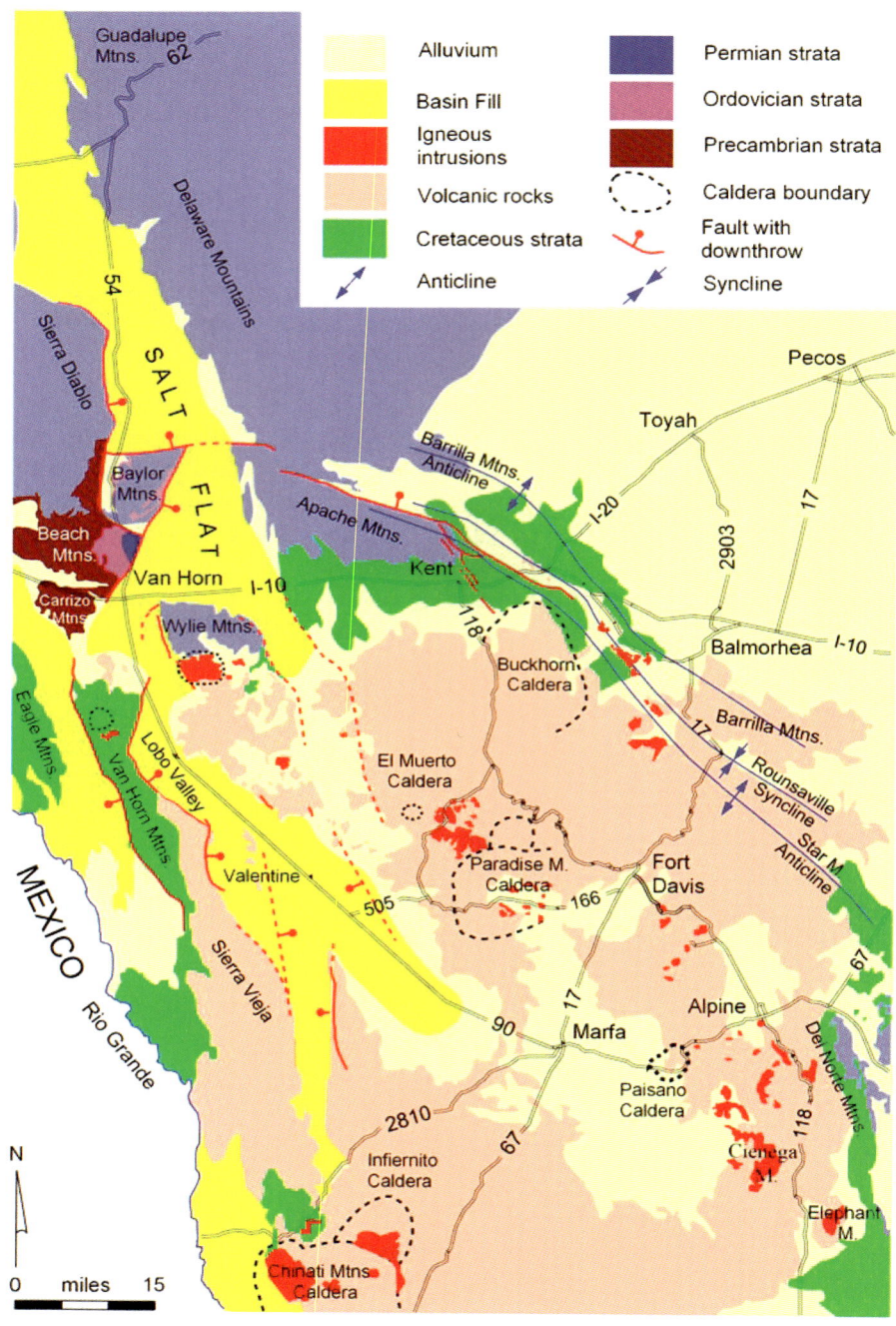

Fig. 2.1: The geology of the Davis Mountains area

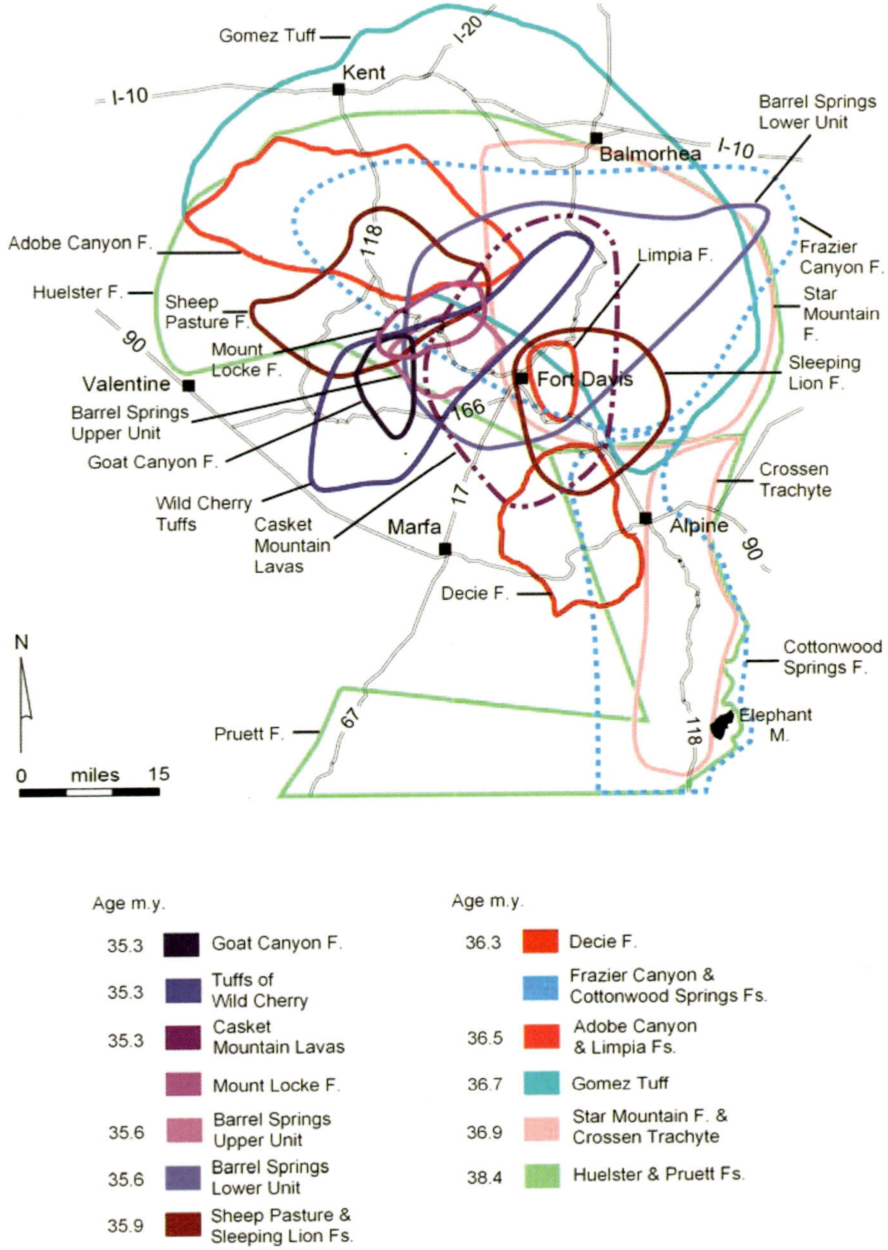

## 2: DEVELOPMENT OF THE MOUNTAINS 15

Fig. 2.2: The outcrop limits of the various volcanic layers making up the Davis Mountains are shown above. Lava layers are shown in shades of red and purple, pyroclastic and volcaniclastic layers in shades of green and blue. Their ages are given in the table. Individual layers are described later in this chapter.

A number of scientific papers have identified factors that contributed to these widespread lava flows. One is that the lavas were hotter than usual when erupted; their temperatures have been calculated by examining the composition of minerals such as pyroxenes. A second is that the lavas may have had a higher than normal content of volatile components. A third is that several of the units originally identified as lavas have turned out to be *rheomorphic* ash-flows, in which the rocks became mobile through at least partial melting.

The igneous rocks of the Davis Mountains are underlain by sedimentary rocks, mostly limestones and sandstones. *Clastic* sedimentary rocks such as sandstone are part of the cycle of nature in which rocks become exposed to the elements of water, wind, and frost, break up, and are carried down as particles or *clasts* into rivers, lakes or oceans where they settle and become cemented into sedimentary rocks over time.

Clastic sedimentary rocks are mostly classified by size of the clasts that make them up. The coarsest, *conglomerate*, can contain clasts ranging from large boulders to very fine-grained clay in the interstices. *Sandstone* is made up of sand-sized grains, usually of the mineral quartz, cemented together by materials such as silica (silicon dioxide), lime (calcium carbonate) and iron oxide. Rocks composed of particles finer than sand are called *shale* if they are layered or have partings and *claystone* or *mudstone* if they do not.

*Limestones* are chemical sedimentary rocks composed mainly of the mineral calcite (calcium carbonate). They can be formed in a wide variety of ways. Certain kinds of miniscule plants called algae, for example, extract calcium carbonate from seawater or fresh water and create limey ooze, which eventually turns into unbedded limestone. Calcium carbonate can precipitate directly from seawater and create similar rocks. Limestone reefs are built by animals such as corals, and in the distant past, by sponges. Impure limestones containing sand or mud are called *sandy limestone* or *marl*, respectively.

Most sedimentary rocks have variations in color or grain size that separate the rock into beds or *strata*, the plural of the Latin word stratum, which means something spread or laid down. Sedimentary rocks without obvious bedding are called *massive*.

*Metamorphic rocks* are sedimentary or igneous rocks that have been altered by heat or pressure or both. For example, hot igneous intrusions can alter rocks that they come into contact with, a process called contact metamorphism. When rocks come under high pressures from tectonic forces, minerals assume new forms; clay turns into mica, for example. Metamorphic rocks underlie the Davis Mountains at depth and crop out around Van Horn.

# 2: DEVELOPMENT OF THE MOUNTAINS

## The Earth's Crustal Plates

The geological history of the western United States, from about Balmorhea westward, is very much bound up with the disappearance of a section of the Earth's crust below western North America.

The Earth's outer layer, the *lithosphere*, is the rigid outermost layer of the Earth. Under the oceans, its average thickness is about 40 miles, under continents, 70 to 90 miles. It is underlain by the much weaker *asthenosphere*, which reacts to stress like a fluid, and extends down to a depth of 280 miles.

The lithosphere is divided into a number of *plates*, 6 large and at least 14 smaller ones that move slowly around the planet at rates of up to several inches per year. New lithosphere is generated at *spreading centers* in mid-ocean ridges such as the mid-Atlantic Ridge. These ridges circumnavigate the entire globe and are commonly 300 to 600 miles wide and from 6,500 to 10,000 feet above neighboring ocean basins. They are made up of segments connected by transform faults in a zigzag fashion. In a transform fault, rocks on either side of the fault slide past each other horizontally. Ridges are shown as solid black lines on Fig. 2.3, the faults as dotted black lines.

Plates are consumed at *subduction zones* (red on Fig. 2.3) where the oldest and coldest plate is subducted under the other at roughly 45 degrees, creating enormous stresses, earthquakes and volcanism as it does so.

The upper part of the lithosphere is called the Earth's crust. It can be oceanic or continental. Oceanic crust underlies the oceans and is 5 miles thick on average. Continental crust, which makes up the continents, is rather less dense than oceanic crust and is 12-40 miles thick, the thicker parts lying under large mountain ranges such as the Andes. Under the Davis Mountains, for example, the crust is about 28 miles thick.

Most plates are made up of both oceanic and continental crust. The North American plate includes North America plus the western half of the north Atlantic (Fig. 2.3). On its east, it meets the Eurasian and African plates at the mid-Atlantic ridge. On its west, it began overriding the Farallon plate in a subduction zone about 240 million years ago (Ma). At 29 Ma, the trailing edge of this plate, the East Pacific Rise spreading center, collided with the North American plate and subduction ended.

Only fragments of the Farallon plate remain on the surface, the Cocos and Juan de Fuca plates, but the plate can be traced under the North American plate by geophysical seismic methods. Fragments of its trailing edge have been found underlying Arizona, New Mexico, Texas and the Gulf of Mexico at depths from 200 to 400 miles (Fig. 2.4). It has even been detected under the western Atlantic at a depth of 1,700 miles.

## 2: DEVELOPMENT OF THE MOUNTAINS

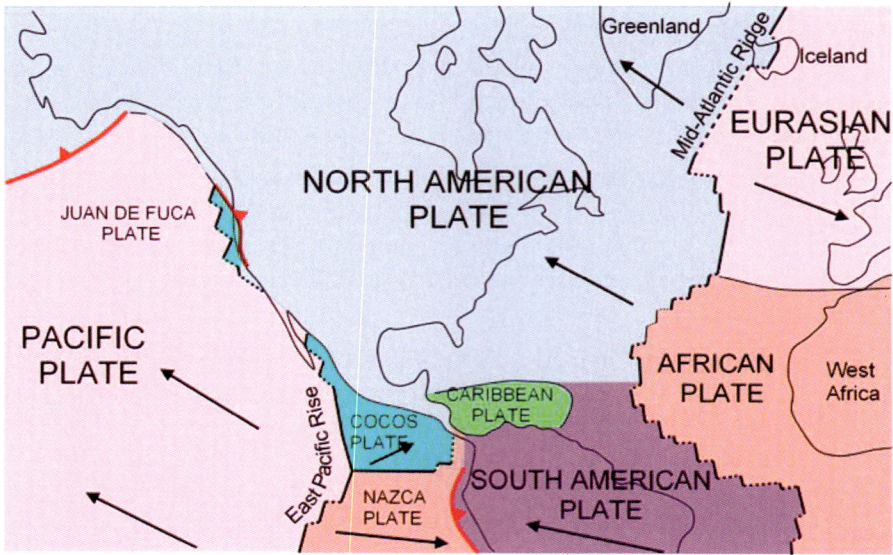

Fig. 2.3: The North American plate is bounded on the west by the Juan de Fuca and Cocos plates, remnants of the Farallon plate, and the Pacific plate. The latter intersected the North American plate about 29 Ma. The red lines are present-day subduction zones with barbs showing the direction of subduction. Black arrowed lines show the direction of plate travel today (adapted from Windley).

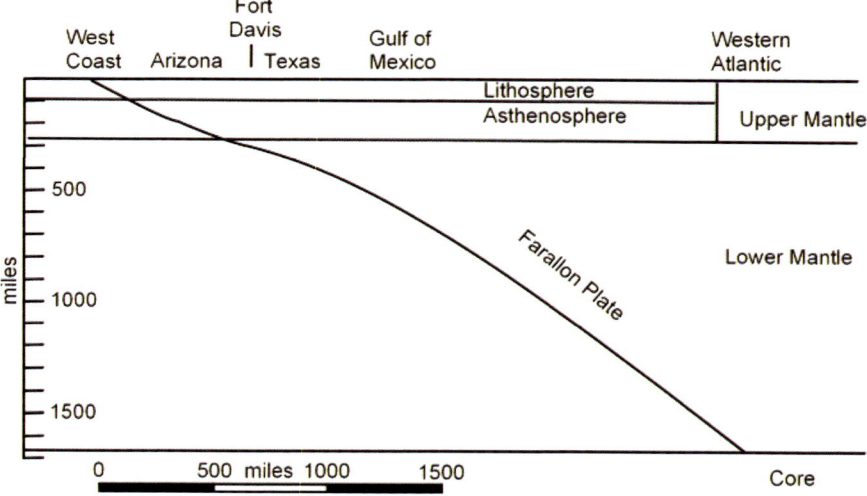

Fig. 2.4: Current position of Farallon plate from geophysical seismic data. Fragments of the plate underlie Arizona, east Texas and the Gulf of Mexico at depths of 200 to 400 miles and the western Atlantic at 1,700 miles (data from Baldridge).

# 2: DEVELOPMENT OF THE MOUNTAINS

| Eon | Era | Period | Epoch | From m.y. | To m.y. |
|---|---|---|---|---|---|
| Phanerozoic | Cenozoic | Quaternary | | 1.6 | 0.0 |
| | | Tertiary | Pliocene | 5.3 | 1.6 |
| | | | Miocene | 23.8 | 5.3 |
| | | | Oligocene | 33.7 | 23.8 |
| | | | Eocene | 54.8 | 33.7 |
| | | | Paleocene | 65.0 | 54.8 |
| | Mesozoic | Cretaceous | Upper Cretaceous | 98 | 65 |
| | | | Lower Cretaceous | 144 | 98 |
| | | Jurassic | | 206 | 144 |
| | | Triassic | | 251 | 206 |
| | Paleozoic | Permian | | 290 | 251 |
| | | Pennsylvanian | | 323 | 290 |
| | | Mississippian | | 354 | 323 |
| | | Devonian | | 417 | 354 |
| | | Silurian | | 443 | 417 |
| | | Ordovician | | 490 | 443 |
| | | Cambrian | | 543 | 490 |
| Proterozoic | Neo- | | | 1,000 | 543 |
| | Meso- | | | 1,600 | 1,000 |
| | Paleo- | | | 2,500 | 1,600 |
| Archean | | | | 4,600? | 2,500 |

Fig. 2.5: Geological time scale; "m.y." is an abbreviation for "million years".

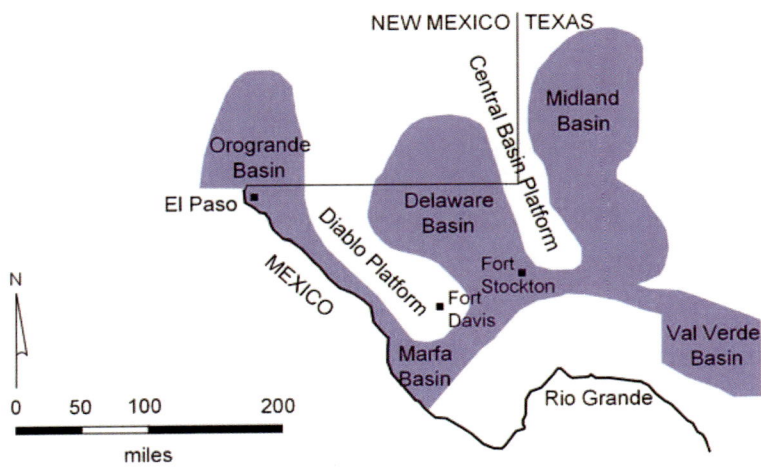

Fig. 2.6: Uplifts and ocean basins (blue), collectively known as the Permian Basin.

## Geologic Time Scale

The geological time scale begins with the creation of the Earth, about 4.6 billion years ago and ends at the present day (Fig. 2.5). The basic unit of time is one million years or *m.y.* Events are described as having occurred or taken place so many millions of years ago, abbreviated as *Ma*.

The time scale was developed for sedimentary rocks and the major divisions are based on events in the fossil record. For example, at the beginning of the Paleozoic era (from the Latin for *early life*), the number of species in existence expanded hugely and sea life developed shells that could be preserved in sediments, so large numbers of fossils occur in Paleozoic rocks. The Paleozoic-Mesozoic *(early life-middle life)* boundary is marked by a massive extinction of life in which 90 to 95 per cent of marine species and perhaps 75 per cent of terrestrial species disappeared. Another mass extinction took place at the Mesozoic-Cenozoic *(middle life-recent life)* boundary, when dinosaurs disappeared and mammals took over as the dominant life form on land.

The remainder of the chapter goes through the history of the Davis Mountains area from the oldest rocks found there to the end of the main period of volcanism. Subsequent history is recounted in Chapter 9.

## Proterozoic Rocks

In the Archean and Proterozoic eons, the North American continent was assembled from the north southwards from sections of continental crust being attached at subduction zones. The oldest part of the continent, in Canada, is over 2½ billion years old.

In the Davis Mountains area, a section of crust 1,100 m.y. old attached itself to the southeast coast of North America at about 1,000 Ma, extending the continent to the southeast in an event called the Grenville Orogeny. In the orogeny, the Carrizo Mountain Group, a series of sandstones, shales, limestones and rhyolites, 19,000 feet thick and 1,350 million years old, was metamorphosed by heat and pressure into quartzite, slate, schist and marble, and thrust over a younger set of limestones and lavas, the Allamoore Formation, 1,250 million years old.

The metamorphosed rocks are exposed in road cuts from 3 to 9½ miles west of Van Horn on Interstate 10 and can be seen in the Carrizo Mountains, just west of Highway 90. They are intriguing for their astonishingly fresh appearance and sharp boundaries between quartzite layers.

## Paleozoic Rocks

After a long period of time of which no record remains, shallow seas invaded the area at the beginning of the Paleozoic era. Thin bedded sandstones of the Cambrian Bliss Sandstone, the earliest Paleozoic formation exposed, crop out at the base of the Baylor Mountains. They were followed by Ordovician to middle Devonian limestones that are now exposed in the Beach and Baylor Mountains north of Van Horn.

An uplift, the Diablo Platform (Fig. 2.6), developed northwest-southeast across the area of today's Davis Mountains and was largely above water during the early Paleozoic. To its north, a large ocean basin existed from the late Cambrian to the late Mississippian, the Tobosa Basin.

During the late Mississippian and early Pennsylvanian periods, the Tobosa basin was transformed into several basins separated by uplifts, collectively called the Permian Basin. The Diablo Platform continued rising and largely remained above water.

Reefs developed along the shores of the Delaware Basin north of Fort Davis, the most spectacular of which is the Capitan Reef, now in view at the Guadalupe Mountains (the Carlsbad Caverns are in the Capitan Reef), the Apache Mountains just north of Interstate 10, and the Glass Mountains east of Alpine (Fig. 10.3).

At the end of the Permian period, the ocean receded and the basins dried up. As they did so, minerals in the seawater crystallized out to form a salt and gypsum cap over older Permian sedimentary rocks providing a seal that later trapped vast accumulations of oil and gas.

A deep water basin, the Chihuahua Trough, part of a rift stretching from south-central Arizona to the Gulf of Mexico, developed in the Middle Jurassic roughly along the present Rio Grande valley from El Paso to Presidio. Through the remainder of the Jurassic and the early Cretaceous, the Trough filled up with evaporites followed by limestones, 25,000 feet thick.

## Cretaceous Rocks

The ocean receded from the Davis Mountains at the end of the Permian period and did not return until 150 million years later, about 100 Ma in the mid-Cretaceous, by which time the area had been leveled by erosion.

The Cretaceous period was a time of high sea levels throughout the world caused by warm temperatures that melted the polar ice caps and released water to the oceans, and active spreading centers that were wider and taller than less active ones and displaced more water than less active ones.

## 2: DEVELOPMENT OF THE MOUNTAINS

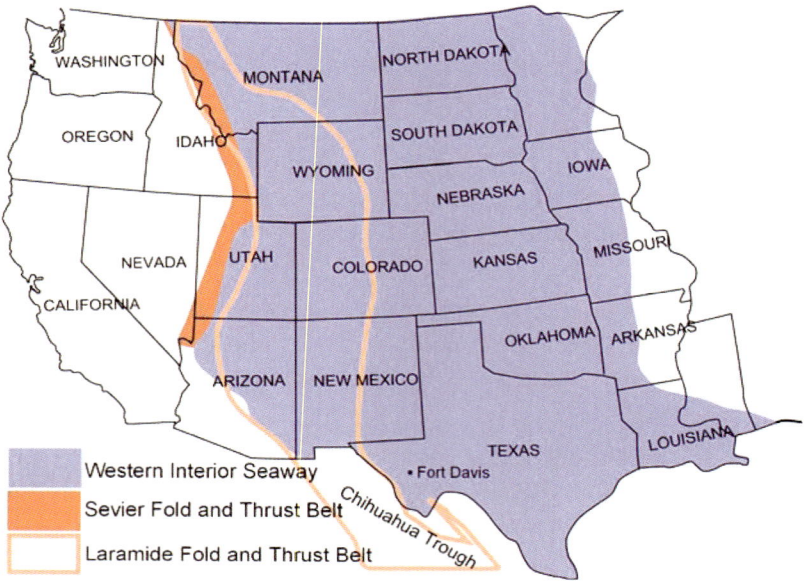

Fig. 2.7: The Cretaceous Western Interior Seaway covered all Texas by 100 Ma and retreated west to east across the Big Bend at 78 Ma.

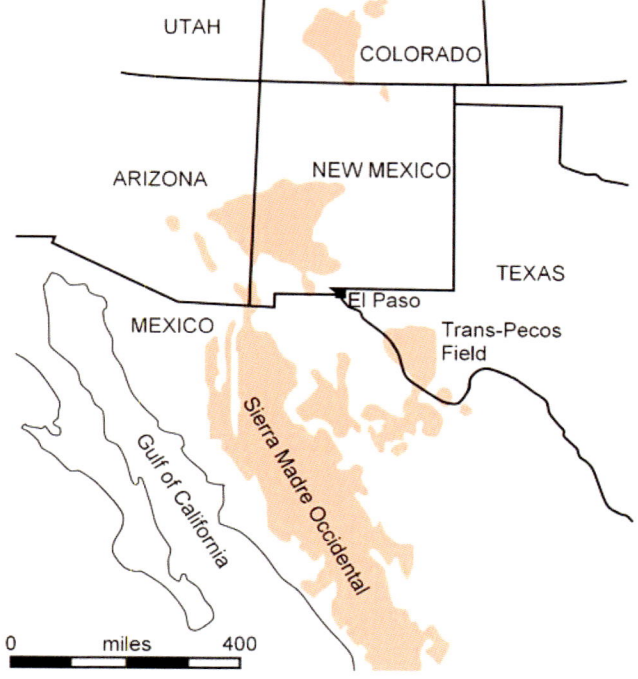

Fig. 2.8: The Trans-Pecos is the most easterly of the Eocene to Late Oligocene felsic volcanic fields in North America.

In western North America, subduction of the Farallon plate which had begun during the Triassic period at about 240 Ma resulted in a nearly-continuous series of orogenies, the processes by which fold belt and other mountainous areas are created. Staggering stresses come from subduction. Imagine a plate 40 miles thick, 1,000 miles long and several hundred miles wide being jammed into the asthenosphere.

Subducted plates are far from smooth and from time to time get caught up in upper plates. The world was alerted to the enormous power of these forces in December, 2004 when, in a subduction zone offshore Sumatra, part of an upper plate that had been caught by a subducting plate suddenly broke free, rising up an average of 33 feet along a 800-mile zone, displacing the ocean and creating a terrifying tsunami that killed more than 100,000 people along the shores of the Indian Ocean. The entire Earth oscillated like a bell for weeks afterwards with ground movement of as much as 0.4 inch everywhere on its surface.

Early orogenies only affected the western United States but they moved east with the leading edge of the Farallon plate and in the Cretaceous period, two orogenies directly affected the Davis Mountains area. In the first, the Sevier Orogeny, which lasted from about 150 to 50 Ma, a basin developed in the central continent along the eastern margin of the orogeny's fold and thrust belt which ran through Idaho and Utah. The ocean gradually encroached into this basin from the Gulf of Mexico from about 150 Ma. It covered all Texas by 90 Ma and at its peak, stretched from the Gulf of Mexico to Alaska (Fig. 2.7). The sea was warm and relatively shallow, 1,300 feet at maximum. It was able to take a great deal of calcium carbonate into solution from the Permian landscape it washed over. Much of the calcium carbonate was redeposited as limestone in west Texas, partly by action of algae, partly by precipitation, which travelers now see along the highways from Austin to Van Horn.

Towards the end of the Cretaceous period, the ocean level gradually declined and the waters became muddy. Upper Cretaceous rocks in west Texas are mainly soft mixtures of marl and mudstone that have only survived around the Davis Mountains where they have been protected by overlying volcanic rocks.

**The Laramide Orogeny**

The second orogeny to affect the Davis Mountains is called the Laramide Orogeny, which lasted from about 80 to 40 Ma. It overlapped and continued the Sevier Orogeny and created a large fold and thrust belt to the east of the Sevier belt from Canada to Chihuahua in Mexico (Fig. 2.7). Both orogenies shortened and thickened the Earth's crust across western North America, creating uplifts with intervening basins.

| Unit | Age m.y. | Description |
|---|---|---|
| Weston & Barillos Domes | 32.8 | Rhyolite intrusions. |
| Mitre Peak intrusions | 34.6 | Rhyolite intrusions. |
| Goat Canyon Formation | 35.3 | Rhyolitic ash-flow tuff, 500 ft. thick. |
| Wild Cherry Formation | | Fine-grained rhyolite tuff, up to 355 ft. thick, produced from Paradise M. Caldera in eastern Davis Mountains. |
| Casket Mountain lavas | | Up to 5 flows of porphyritic rhyolite capping ridges in southern, central and eastern Davis Mountains; 1,600 ft. thick on Blue M.; probably erupted from widely spaced fissures. |
| Mount Locke Formation | undated | Quartz trachyte & rhyolitic porphyry in central Davis Mountains, gray when fresh to brownish gray to reddish brown when weathered; maximum thickness 580 ft. |
| Tuffaceous sediment, mafic lava | | Tuffaceous sediment & mafic lava are at same horizon as Mount Locke Formation in Blue M. and Musquiz Canyon in southeastern Davis Mountains. |
| Barrel Springs Formation | 35.6 | Upper unit: rhyolitic lava. Lower unit: strongly rheomorphic ash-flow tuff or lava in central and eastern Davis Mountains; source presumably buried in central Mountains; up to 500 ft. thick. |
| Sheep Pasture Formation | undated | Rhyolite, grayish purple to brown, in northern and western Davis Mountains; 510 ft. thick on Sheep Pasture M. |
| Sleeping Lion Formation | 35.9 | Single rhyolite flow in central & southeast Davis Mtns., probably from source northwest of Fort Davis; 630 ft. thick at maximum. |
| Moore Tuff | | Rhyolitic ash-flow tuff erupted from El Muerto caldera; up to 800 ft. thick in caldera, decreasing to 150 ft. 1½ miles outside caldera. |
| Decie Formation | 36.3 | Rhyolite and quartz trachyte lavas with interbedded tuffs; source multiple fissures 10 miles west of Alpine; total thickness up to 3,000 ft. |
| Frazier Canyon Formation Cottonwood Springs Formation | undated | Tuff and tuffaceous sandstone with lenses of pebble conglomerate; Frazier Canyon reaches maximum thickness of 1,150 ft. in eastern Davis Mountains. Interbedded mafic lava flows pinch out to north and increase in number to south; up to 1,200 ft. thick near Alpine, thinning northwards; occur from 30 miles south of Alpine to 5 miles north of Fort Davis. |
| Adobe Canyon Formation | 36.5 | Two or possibly three rhyolite flows, each 400-850 ft. thick, in north central & northwestern Davis Mountains. |
| Limpia Formation | | Quartz trachyte lavas overlying Gomez Tuff east & southeast of Fort Davis. |

## 2: DEVELOPMENT OF THE MOUNTAINS

| Unit | Age m.y. | Description |
|---|---|---|
| Gomez Tuff | 36.7 | Rhyolitic ash-flow tuff in northern & eastern Davis Mountains; source is Buckhorn Caldera; up to 2,000 ft. thick in caldera, abruptly thinning to a maximum of 480 ft. outside the caldera. |
| Star Mountain Formation | 36.8 | Multiple rhyolite to quartz trachyte lava flows in eastern Davis Mountains; individual flows 200-600 ft. thick; one source identified near Balmorhea. |
| Crossen Trachyte | | Two lava flows of porphyritic rhyolite to quartz trachyte from Musquiz Canyon to Elephant M.; grayish- to reddish-brown, weathers to rusty brown with pitted surface; maximum thickness 265 ft.; source probably multiple fissures buried by flows. |
| Rhyolite domes | 36.9 | Rhyolite domes scattered throughout northeastern Davis Mountains. |
| Cherry Canyon intrusion | | Cherry Canyon rhyolite intrusion emplaced in Buckhorn Caldera prior to eruption of Gomez Tuff. |
| Huelster and Pruett Formations | 38.4 | Reworked tuffs; mafic lavas near base of Huelster Formation in N.E. Davis Mountains. |

Fig. 2.9: Summary of volcanic activity in Davis Mountains (adapted from Henry, Kunk & McIntosh).

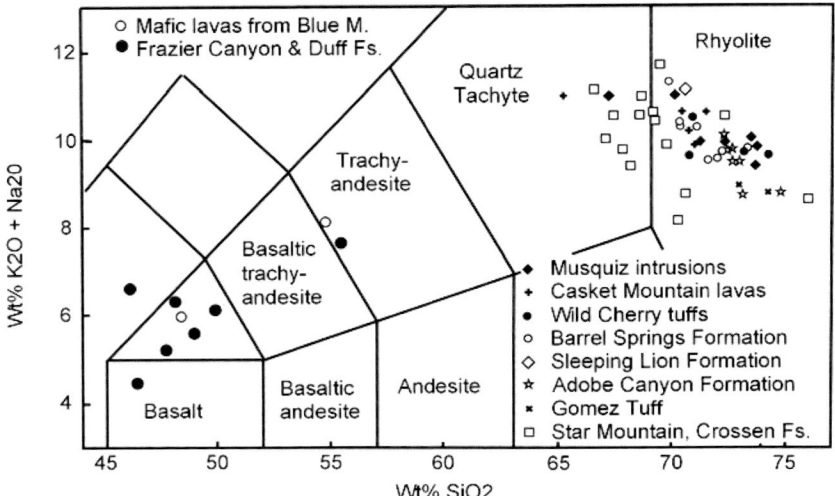

Fig. 2.10: A large majority of Davis Mountains volcanic rocks have a chemical composition in the quartz trachyte and rhyolite groups. Mafic lavas, when they occur are slightly more alkaline than basalt (from Henry, Kunk & McIntosh).

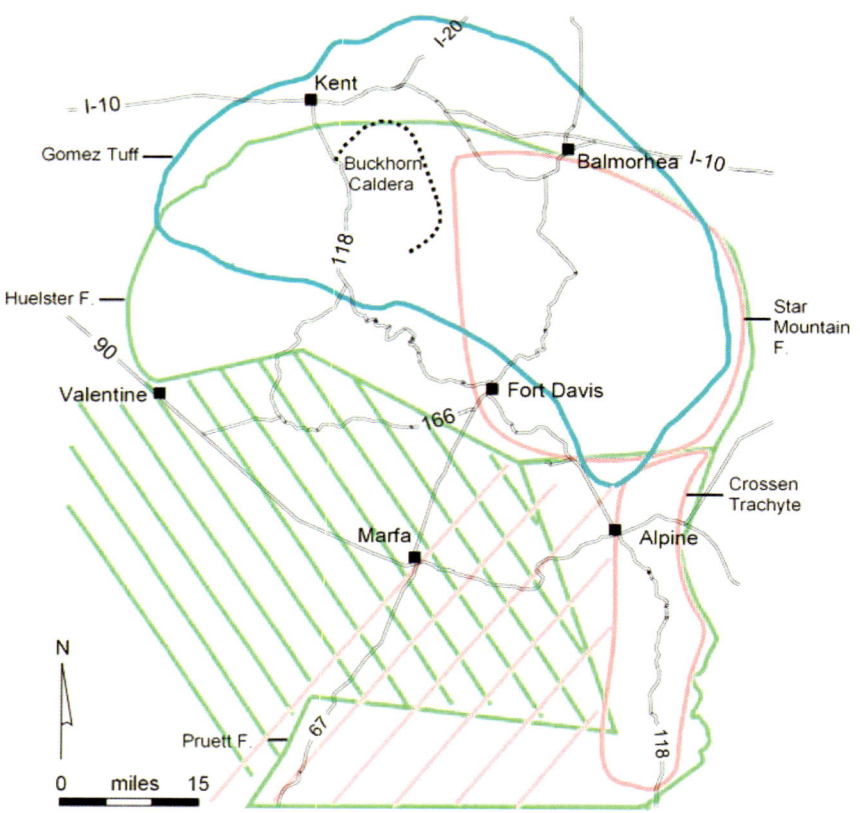

Fig. 2.11: The Huelster and Pruett pyroclastic flows were the first to appear in the Davis Mountains. They probably extend to the west, as shown by the cross-hatching, but are hidden by overlying strata and alluvium. They were followed by the Star Mountain and Crossen Trachyte lavas, the first lavas to appear in the Davis Mountains. The Crossen unit becomes a trachyte in the south, hence the name. It probably extends to the west under overlying formations, too, as shown by its cross-hatching. The widespread Gomez Tuff overlies the Star Mountain lava in the east. The missing sector of the Buckhorn caldera is buried under overlying Adobe Canyon lavas.

# 2: DEVELOPMENT OF THE MOUNTAINS

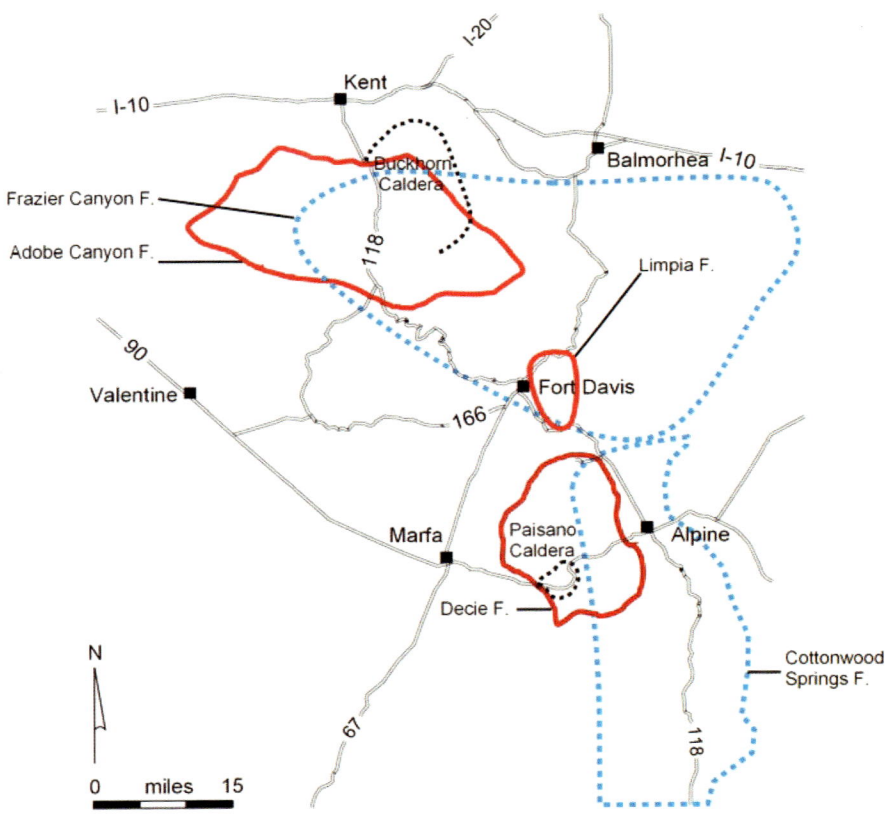

Fig. 2.12: Adobe Canyon rhyolite flows overlie the Gomez Tuff in the northwest. The area covered by the Limpia Formation is more typical of felsic lava bodies. The Frazier Canyon Formation and the similar Cottonwood Springs Formation are broadly contemporaneous, and consist of varying proportions of tuffs and mafic lavas. The Decie Formation was extruded from fissures in the Paisano volcano. The small Paisano caldera developed during the million years the volcanic center was active.

The effects of the Laramide Orogeny were uneven. On the Colorado Plateau (Fig. 9.1), it had only minor effects, but further east it created the high mountains of the Front Range in Colorado and the Sangre de Christo Mountains of Colorado and New Mexico. In the Big Bend area, it created a fold belt parallel to the Rio Grande over the now-filled Chihuahua Trough, the mountains you see across the Rio Grande from Presidio. It also created the Del Norte-Santiago-Sierra del Carmen mountain range running south from east of Alpine into Mexico.

The orogeny ended about 40 Ma but much of the uplifted section of the continent is still mountainous today, partly as a result of the Laramide Orogeny and partly the result of later events (see Chapter 9).

## Volcanic Eruptions

Powerful volcanic eruptions began in the southwest United States and Mexico not long after the end of the Laramide Orogeny (Fig. 2.8). The Davis Mountains are part of the Trans-Pecos field, on the very eastern edge of the activity, which also includes the Chinati and Chisos volcanic centers. The igneous activity was undoubtedly related to the subducting Farallon slab although the details are the subject of intense debate. Volcanic zones above subducting plates are common; volcanic activity related to subduction of the Cocos slab continues today in central Mexico.

The volcanic sequence in the Davis Mountains consists of a series of lava flows separated by widespread pyroclastic and volcaniclastic strata. In a paper published in 1994, Henry, Kunk and McIntosh identified 6 major lava episodes that occurred between 36.8 and 35.3 Ma at roughly 300,000-year intervals. Their work was based on a significantly more accurate method of isotopic dating than had been used in prior studies of Davis Mountains rocks and so gave rise to some revised correlations. The revised sequence and dating is given in Fig. 2.9. The range of chemical compositions is given in Fig. 2.10. The approximate extents of the lavas and the intervening pyroclastic formations are given in Figs. 2.11-2.15. Lava units are shown in shades of red and purple, pyroclastic units in shades of blue and green.

## The Huelster and Pruett Formations

Volcanic activity throughout the Big Bend began with pyroclastic flows (Fig. 2.11). In the Davis Mountains, the basal formation is called the Huelster Formation. It consists of a series of ash-fall tuffs with conglomerate, sandstone and non-marine limestone lenses and includes some interbedded lava flows. The formation is 250 feet thick in the northern Davis Mountains, thickening eastwards to 500 feet in Cherry and

Madera Canyons and about 400 feet in the Barilla Mountains. Its source is unknown but the lava flows indicate that there were at least some local eruptions. A conglomerate of limestone and quartzite boulders, the Jeff Conglomerate, occurs at its base.

South of Musquiz Canyon, the basal tuff is called the Pruett Formation. Chemical analysis shows it to be from a different source than the Huelster but it is otherwise very similar. The western extent of these formations is unknown but in the Sierra Vieja, the lowest volcanic formation is a tuffaceous sandstone with the Jeff Conglomerate at its base so they may continue under alluvium and basin fill as shown by the diagonal lines on Fig. 2.11.

## Star Mountain and Crossen Lavas

The Huelster Formation was followed by the first of the lava cycles identified by Henry, Kunk and McIntosh. The cycle produced two widespread rhyolitic and trachytic lavas, the Crossen Trachyte south of Mitre Peak, and the Star Mountain Formation in the Davis Mountains (Fig. 2.11). A third unit, the Bracks Rhyolite in the Sierra Vieja west of the Davis Mountains, is about the same age, 36.8 Ma. In the Star Mountain unit, flows from 130 to 825 feet thick can be traced for as much as 12 miles. Up to three flows can occur in any one place and a total of six flows have been identified.

The Crossen unit consists of at least two flows, one extending for some 32 miles south of Alpine. The second smaller flow is near the southern end of the formation outcrop. The flows are considered to have come from multiple fissures and volcanic vents now buried under younger strata. Christopher Henry has positively identified one fissure north of Alpine, where a dike of Star Mountain rhyolite passes into an overlying flow on Last Chance Mesa.

## The Gomez Tuff

A hundred thousand years after the creation of the Star Mountain Formation, the Gomez Tuff spread across the northeastern Davis Mountains (Fig. 2.11). One of the most extensive units in the sequence, 70 miles across, it emanated from an area in the north-central mountains which collapsed to create the Buckhorn caldera. This caldera, the largest so far found in the Davis Mountains, is 14 miles north-south by 7½ miles east-west. The tuff thickens abruptly from 500 feet to 2,000 feet at the perimeter of the caldera. It thins away from the caldera to as little as 6 feet in the eastern mountains.

30     2: DEVELOPMENT OF THE MOUNTAINS

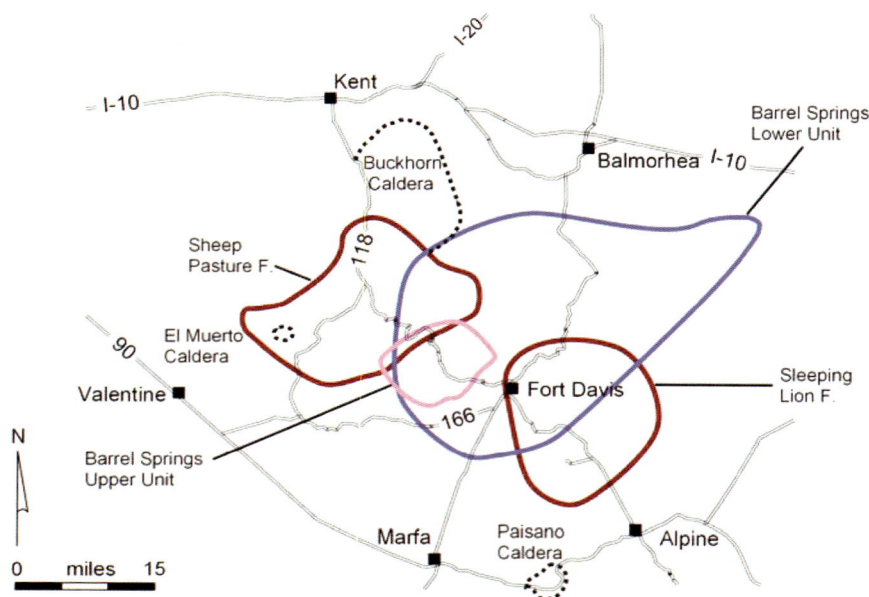

Fig. 2.13: Sheep Pasture and Sleeping Lion lavas followed Frazier Canyon tuffs and mafic lavas and were followed by two Barrel Springs flows, the lower unit a rheomorphic tuff and lava, the upper unit a lava.

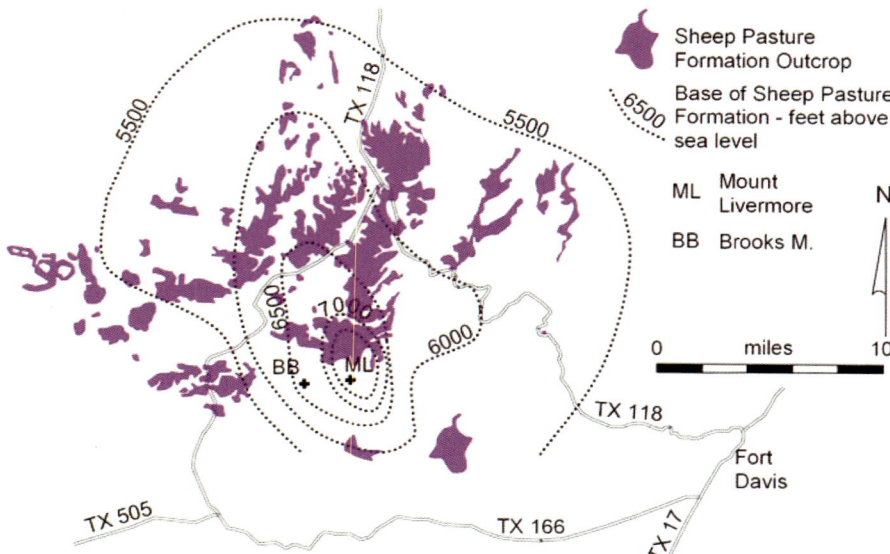

Fig. 2.14: The base of the Sheep Pasture Formation has been domed around Brooks Mountain and Mount Livermore, presumably when intrusions that created the volcanic domes at their summits were emplaced.

## 2: DEVELOPMENT OF THE MOUNTAINS

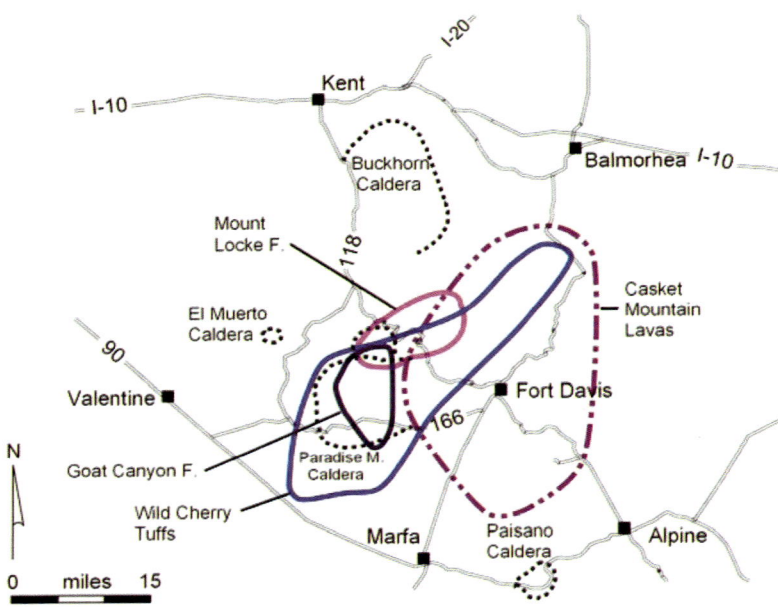

Fig. 2.15: In the final major volcanic cycle, the Mount Locke lava erupted in the central mountains after the Barrel Springs Formation, followed by the Cherry Canyon tuffs from the Paradise Mountain caldera and finally the Casket Mountain and Goat Canyon lavas.

## The Adobe Canyon and Limpia Formations

The Gomez Tuff was followed 200,000 years later by Adobe Canyon rhyolites in the north and northwestern Davis Mountains, and Limpia quartz trachytes south and east of Fort Davis (Fig. 2.12). The former crops out over an area of nearly 500 square miles, extending 35 miles from east to west. It consists of up to three flows with a maximum thickness in excess of 1,000 feet. It probably came from more than one source. The area covered by the Limpia Formation is more typical of felsic lavas, 6 miles by 10.

## Frazier Canyon and Cottonwood Springs Formations

The Adobe Canyon and Limpia lavas were followed in the Davis Mountains by tuffs and mafic lavas of the Frazier Canyon Formation (Fig. 2.12). The age of this formation has not been established except that it is older than the Sleeping Lion Formation and younger than the Gomez Tuff. It is up to 1,150 feet thick in the eastern mountains but is only a few feet thick over much of the Davis Mountains. The lavas die out to the north.

In the south, a similar set of strata occurs at roughly the same level in the volcanic sequence. It belongs to the Cottonwood Springs Formation (Fig 2.12), mafic lavas with interbedded tuffs that overlie the Crossen unit. Six hundred feet of the formation were intersected in an Alpine well and 800 feet in a well north of town. It is the principal source of water for Alpine.

## Decie Formation

This formation, a series of rhyolitic and trachytic lavas with some interbedded tuffs, erupted from the Paisano volcanic center between Alpine and Marfa around 36.3 Ma (Fig. 2.12), 200,000 years after the nearby Limpia Formation. Eruption was through a swarm of fissures about 15 miles in diameter around Paisano Peak. A shallow caldera 3 miles in diameter formed just west of Paisano Peak during eruption. The unit was originally mapped as part of the Duff Formation, a set of tuffs found south of Alpine and thought to have originated in the Chinati Mountains volcanic center.

## Sleeping Lion and Sheep Pasture Formations

The Sleeping Lion rhyolite (Fig. 2.13), a single flow up to 630 feet thick, erupted from a source somewhere northwest of Fort Davis about 35.9 Ma. The flow has created very distinctive palisades along Musquiz Canyon south of Fort Davis and above the Barillos Dome intrusion. It overlaps with Decie Formation strata around Mitre Peak.

The Sheep Pasture rhyolitic lava (Fig. 2.13), found over a wide area in the western Davis Mountains, is very similar to the Adobe Canyon Formation which it overlies in places. It has never been dated but is younger than the Frazier Canyon and older than the Barrel Springs Formations. The unit is about 550 feet thick on Sheep Pasture Mountain. Its source is unknown.

Lava flows of the Sheep Pasture and the overlying Barrel Springs Formations have been domed as much as 2,000 feet on Mount Livermore (Fig. 2.14). Presumably the trachytes that form the Mount Livermore and Brooks Mountain summits are the visible evidence of intrusions that domed the older formations as they pushed up from below.

The Moore Tuff, a small ash-flow tuff not shown in Fig. 2.13, erupted from the El Muerto caldera northeast of Valentine about the same time as the Sleeping Lion Formation.

## Barrel Springs Formation

The Barrel Springs Formation (Fig. 2.13), as redefined by Henry, Kunk and McIntosh by their revised isotopic dating and field mapping, consists of two units. The upper unit, found over a limited area in the southern mountains, is a lava. The lower unit is controversial: Henry, Kunk and McIntosh describe it as either a lava or a strongly rheomorphic tuff, i.e. one that began to flow. An earlier paper by Henry, Parker, Price and Wolff, however, describes the Barrel Springs Formation in the Davis Mountains State Park as transitioning upward from a typical ash-flow tuff at base to a strongly rheomorphic ash-flow tuff. Exposures in the Rounsaville syncline in the northeastern Davis Mountains, correlated by Parker in the same paper with the rheomorphic tuff in the state park, are clearly of ash-flow tuff.

Both units are very similar in chemical composition and age, 35.6 Ma, and probably came from the same unidentified source. The lower unit is separated from the Sleeping Lion Formation by about 30 feet of volcaniclastic tuff, created during the 300,000 year interval between the formations.

## Mount Locke Formation

Undated porphyritic trachytes of the Mount Locke Formation are found between the Barrel Springs lower unit and the tuffs of Wild Cherry in the central mountains (Fig. 2.15). Volcaniclastic strata and mafic lavas appear at the same horizon in Musquiz Canyon and on Blue Mountain.

## Wild Cherry, Casket Mountain and Goat Canyon Formations

Three hundred thousand years after the Barrel Springs Formation, three separate volcanic events took place in the Davis Mountains (Fig. 2.15). Wild Cherry rhyolitic tuffs were produced by the Paradise Mountain caldera. They are 350 feet thick in the caldera, but thin to 200 feet on Blue Mountain and to only 6 feet in the Barilla Mountains. On Pine Peak, they overlie Mount Locke trachyte.

A porphyritic rhyolite found on Casket Mountain and elsewhere, capping many of the ridges and summits in the southern, central and eastern Davis Mountains is of the same age. Five flows, 1,600 feet thick, have been identified on Blue Mountain, the oldest below Wild Cherry tuff, the others above. The flows were probably erupted from a series of unconnected fissures and so are shown as a dashed line on Fig. 2.15.

The third event is the eruption of the Goat Canyon rhyolitic ash-flow tuff from an unknown source. It has an identical age, 35.3 Ma, to the other units. It is 480 feet thick on Pine Peak, overlying Wild Cherry tuff and underlying a rhyolitic lava.

## Younger Volcanic Rocks

Younger undated lavas are found in places in the Davis Mountains, such as the lava that overlies Goat Canyon tuff on Pine Peak. Others include the Puertacitas Formation, basalts and tuffs up to 1,250 feet thick that crop out from Mitre Peak north of Alpine to the Puertacitas Mountains. They came from sources in the Haystacks-Puertacitas Mountains area and lie above and so are younger than the Decie Formation.

In the central Davis Mountains, the Brooks Mountain Formation is a porphyritic trachyte up to 985 feet thick on Brooks Mountain just west of Mount Livermore. The white cap on Mount Livermore, the highest point in the mountains, was originally identified by Anderson as an andesite intrusion, but more recently, Henry, Price and Parker called it as an eroded trachyte dome complex. Both it and the Brooks Mountain trachyte should probably be classified as volcanic domes, created by highly viscous lava squeezed out from volcanic vents late in the eruptive cycle that formed domes above the vents. Neither has been dated.

The Sheep Pasture and Barrel Springs Formation lavas in the Brooks Mountain-Mount Livermore area have been elevated as much as 2,000 feet above their position elsewhere in the mountains. The most likely explanation is that the volcanic dome trachytes on the high mountains are the surface expressions of larger intrusions at depth.

## Igneous Intrusions

Igneous intrusions are exposed around Alpine and on the northern fringes of the mountains west of Balmorhea, where younger volcanic rocks have been eroded away. Several large intrusions are found in the Mount Livermore area, including Sawtooth Mountain, a laccolith. Henry, Kunk and McIntosh dated two laccoliths along Musquiz Canyon at 32.8 Ma, 2½ million after the main volcanic activity ended. Another small intrusion near Mitre Peak was dated at 34.6 Ma. Numerous small syenite outcrops, the Point of Rocks and others, are found in the Paradise Mountain caldera. They are believed to be connected at depth to a single intrusion, one that rose up when magma in the caldera was exhausted, a process called resurgence by geologists.

## Subsequent History

Subsequent events, including the development of rifting along the western margin of the mountains, their tilting to the southwest, the widespread eruption of basalt in the rifted terrain, and erosion of the mountains and development of drainage, are described in Chapter 9.

## 3: APPROACHING THE MOUNTAINS 35

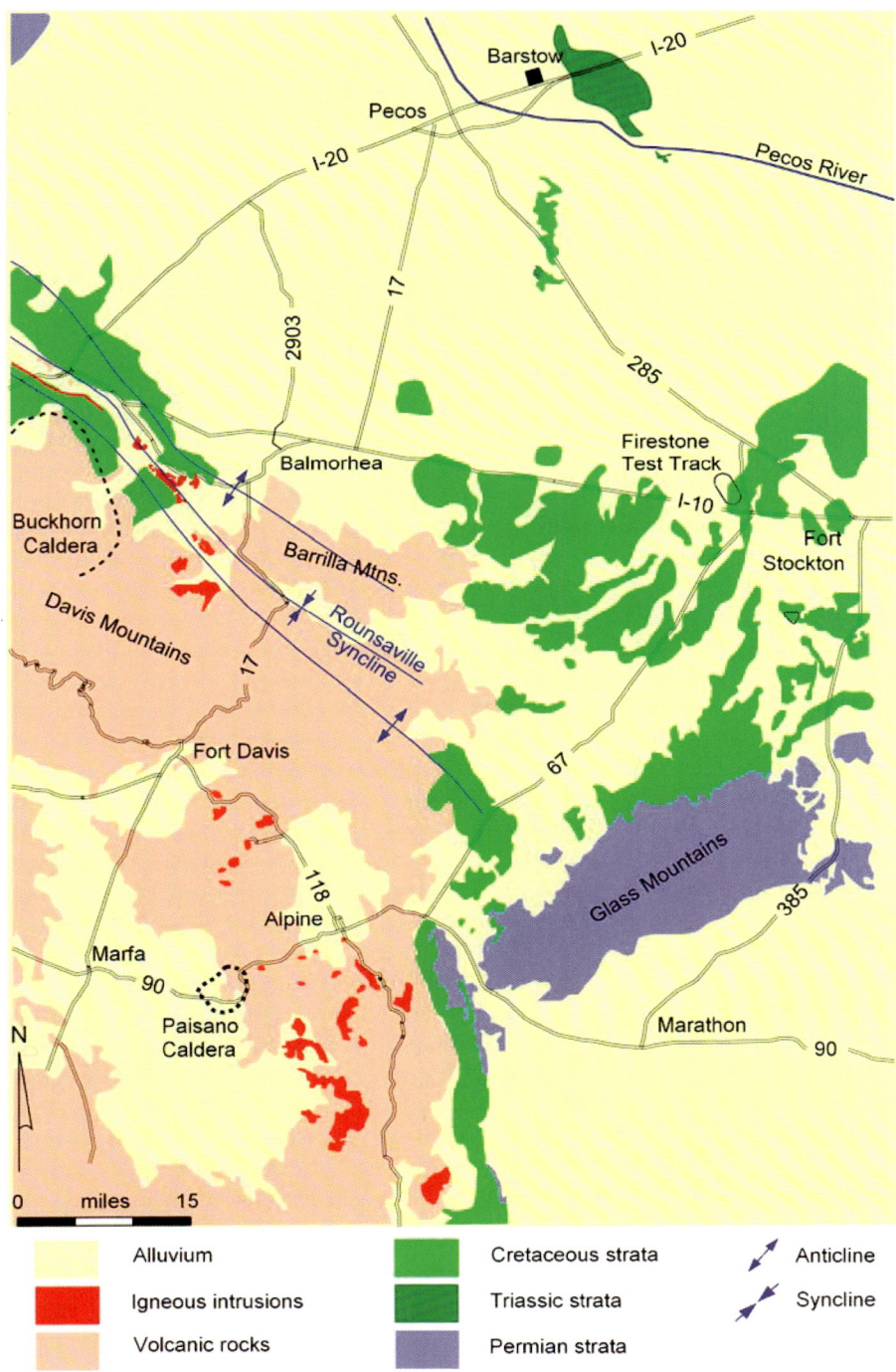

Fig. 3.1: Geology along the approaches to the northeastern Davis Mountains.

## 3: Approaching the Mountains

This chapter gives a brief introduction to the mountains for the majority of visitors who come from the Texas population centers along Interstate 20 or Interstate 10.

Visitors who come south on Highway 62 to Van Horn, turn to page 168 and follow Highway 90 to Fort Davis. Travelers from the east on Highway 90, turn to page 135 and continue on Highway 90 at Alpine.

### Approaching via Interstate 20

Approaching the Davis Mountains along Interstate 20, the highway traverses a terrain of low undulations from Odessa westward, across plains of alluvium, with sand dunes in places, that overlie shales and siltstones of the Triassic Dockum Formation. This non-marine formation is exposed just east of Barstow, where an outcrop crosses the highway (Fig. 3.1). The Davis Mountains come into view shortly before the Barstow exit 10 miles east of Pecos at about 10 o'clock from the crest of a slight ridge, stark and sharp on the horizon.

By the time the traveler arrives at the junction of Highway 17 and Interstate 20 in Pecos, the mountains are on the skyline from about 11:30 to 2 o'clock, not particularly impressive; they are just distant hills. Abandoned farm equipment and buildings are evidence of earlier irrigated cotton farming along much of Highway 17 between Pecos and Interstate 10. The Toyah Basin, which includes most of Reeves County between the Davis Mountains and the Pecos River, is underlain by water-bearing sediments up to 1,500 feet thick. From 1945 onwards, irrigation pumping produced water for cotton and other crops. However, a combination of lower prices for cotton, increased pumping costs caused by higher natural gas prices, and a lowered water table has driven cotton farmers out of business along the highway.

# 3: APPROACHING THE MOUNTAINS

Pumping continues for cattle feed lots, however, and for irrigated crops in the western part of Reeves County. Current water production from the Toyah Basin is around 150,000 acre feet, 49 billion gallons per year, according to the Texas Water Development Board, more than from any other aquifer in Trans-Pecos Texas. Groundwater is replenished by rainfall in the Davis and Barrilla Mountains, the Rustler Hills and the Sacramento Mountains in New Mexico, and by underground water flowing east through the Apache Mountains from the Salt Flat.

At Hoban, nineteen miles from Pecos, Capitol Aggregates of San Antonio operates a sizeable sand and gravel quarry, shipping out material on the Pecos Valley Southern railroad to the Union Pacific line in Pecos. The Pecos Valley line was completed to Toyahvale in 1911; the section between Hoban and Toyahvale was abandoned in 1971.

The mountains are very prominent on the skyline at 3 o'clock by the time you arrive at Saragosa, the only settlement of any size on Highway 17. More than 80 per cent of the town and all its public buildings were destroyed by a tornado in 1987, in which 30 people were killed and 162 injured. Interstate 10 is 2.3 miles beyond the Saragosa turnoff. This road guide continues on page 40.

## Approaching via Interstate 10

From San Antonio west, Interstate 10 crosses the Edwards Plateau and the Stockton Plateau, as the Edwards Plateau is called west of the Pecos River. The highway is on Lower Cretaceous strata, mostly limestones, for the 390 miles from San Antonio to Balmorhea, all deposited in the Western Interior Seaway between 100 and 78 Ma (Fig. 2.7). In places the plateau has been eroded down to sandstones of the lowest unit of the Lower Cretaceous, the Glen Rose Formation. For example, bright red Glen Rose sandstones can be seen in road cuts on the left as you descend to the Llano River valley at Junction. Other deeply eroded sections are around Kerrville and Sheffield, where the Guadalupe and Pecos Rivers have cut canyons up to 500 feet deep down to the lower Cretaceous.

Otherwise, the highway is bordered by road cuts of light gray fine grained limestone, well-bedded in places, with occasional local brown beds of sandy or clayey limestone. The limestone is predominantly a biomicrite, composed of shell and skeleton debris from sea creatures (bio) and limestone mud (micrite) from algae, deposited in the shallow sea. The brown beds come from fine sand or clay carried out to sea from land by rivers or streams.

Erosion diminishes west of the Pecos River and the canyons broaden into wide draws seldom more than 200 feet deep. Shortly beyond Fort Stockton, they die away into rolling countryside with scattered mesas.

| Period | Age m.y. | Group | Description |
|---|---|---|---|
| Lower Cretaceous | 144-98 | Washita Group | Buda Limestone: fine grained, hard, massive, poorly bedded to nodular limestone, light gray to orange, weathers dark gray to brown; 140 feet thick |
| | | | Del Rio Clay: marl, light gray to yellowish gray; changes to San Martine member of the Boracho Formation near Balmorhea; 30 ft. thick. |
| | | Fredericksburg Group | Limestone, dolomite, chert and minor marl; 135-200 ft. thick. |
| | | Trinity Group | Glen Rose Formation: sandstone, fine to coarse grained, becoming conglomeratic northwards; up to 100 ft. thick. |

Fig. 3.2: Lower Cretaceous strata in the Fort Stockton area; a "group" is a combination of similar formations. Compare to Fig. 10.2, the succession further west.

The entire area of central and east Texas was tilted up to the west sometime after the area became dry land at 78 Ma. For example, the boundary between the Fredericksburg and Washita groups of the Cretaceous Formation is at 1,950 feet elevation at Junction and 2,850 feet at Fort Stockton (see also page 149). The inclination, about 5 feet per mile, is imperceptible as you drive along, however.

At Fort Stockton we join the route taken by Lieutenant William Whiting and his party in 1849. Whiting had been commanded by army commander General Worth to find out whether there was a route suitable for military supplies between San Antonio and Santa Fe. He assembled a party of 16, including his West Point classmate Lieutenant Smith, a topographical engineer, the scout Dick Howard, and two Mexican drovers, one of which, Jose Policarpo Rodriguez, was also an accomplished tracker and hunter.

Whiting left a diary and journal that give vivid accounts of travel in the Trans-Pecos in 1849, and is the first account of the area written in English. Policarpo's reminiscences have also been recorded, and are in some ways more revealing, as he could speak Spanish to the Apaches and to Mexican slaves, present in every Apache household.

The Whiting party camped at Awache spring, later called Comanche Springs, south of City Hall in present-day Fort Stockton...

> "...a clear gush of water which bursts from the plain, unperceived until the traveler is immediately upon it, and soon swells in a clear, running brook abounding in fresh fish and soft-shell turtles."

The Barrilla Mountains (5,568 feet), and the Davis Mountains behind them, are first seen on the horizon at 10 to 11:30 o'clock from a rise about a mile west of Fort Stockton. From this distance, 40 miles away, they appear as a block, the first of many on the interstate to California. The highest point visible is Star Mountain (6,344 feet), which stands up as a very distinctive pyramid on the horizon midway along the range. The mountain front on the north or right side of the block is quite sharp, dropping 2,300 feet in four miles to the interstate.

## Belding Irrigated Area

A long, gentle incline leads to the Belding Exit at Milepost 253. The irrigation area of Belding is on the left. Farming has been conducted here since at least 1870 when the census included a settlement called Lylesville after George Lyles, who farmed at Leon Springs, just beyond the exit on the left. In 1913, the Kansas City, Mexico and Orient railroad opened a station at Belding, 6 miles south of the springs, named after one of the railway company's directors.

Later, runoff from Leon Springs was dammed to create Leon Lake, and a hotel was moved there from Belding. The springs stopped running in 1958, however, as irrigation at Belding lowered the water level. Comanche Springs at Fort Stockton dried up about the same time. Belding is now the site of a 2,200-acre pecan orchard, which is irrigated by water pumped from 200 to 300 feet underground. Alfalfa, small grains and cotton are also grown in the Belding area.

The interstate for the next several miles crosses a series of low ridges at the summits of which limestones are exposed in road cuts. At 9 o'clock, Twelve-mile Mesa (3,733 feet) is the last large mesa along this highway, standing about 400 feet above the flats below. Down below the road on the left, the extensive area of farmland is in the Belding irrigated area.

The road climbs up from the Highway 67 exit to Alpine and the Big Bend to a summit in which more limestones are exposed in road cuts on either side of the highway. From the summit, the Davis Mountains create a panorama on the horizon from about 10 to 12 o'clock, although still quite a distance away.

## Firestone Texas Proving Ground

Firestone Road is on the right just beyond Milepost 247. The white buildings along it are part of Firestone's 6,000-acre Texas Proving Grounds where tires and cars are tested. The 7.7 mile track opened in 1955 and employs a labor force of about 20. The area was chosen for its mild climate. The average temperature is 67 degrees, annual rainfall averages 12 inches and the sun usually shines 360 days a year so the roads last a long time and test conditions are ideal nearly all year round.

The terrain is fairly level and featureless north of the highway; it tends to low swales to the south. In the springtime, much of this area is covered by a carpet of yellow flowers.

The Glass Mountains on the horizon at 9 o'clock, 30 miles away, are on the northwest side of a dome that rose up around Marathon after the end of the Cretaceous period. Cretaceous strata have been mostly eroded away on the mountains, exposing underlying Permian limestones. The upper part of the ridge is capped in part by a limestone reef, the Capitan Reef, which appears again in the Apache Mountains 70 miles to the west (see page 159).

Three miles ahead, the road cut on the left is in well-bedded Washita limestone, younger than the Fredericksburg strata intersected further east. The highway runs on Washita limestones for the next 20 miles with road cuts occurring at ridge crests.

## Barrilla Mountains

At Mile 236, the highway climbs up a slight incline to one such crest and the Barilla Mountains come into clear view at 11 o'clock. To their left, a tongue of Star Mountain lava, 350 feet thick with Huelster tuffs at base, creates a long low mesa terminating in a sharp escarpment at about 10 o'clock. This is the easternmost exposure of Davis Mountains volcanic rocks.

From the rest area at Mile 233.8, Tucker Hill, a decayed Washita limestone mesa is in the near foreground at 10 o'clock. Several other low mesas can be seen south of the highway in front.

By the Reeves County line at Mile 226.6, a continuous ragged escarpment of Washita limestones capped by a 10-foot limestone bed is on the left. The escarpment ends about 2 miles ahead at which point you can see through to the Barrilla Mountains at roughly 9 o'clock.

Just beyond the Hoefs Road exit at Mile 222 the road descends on to the flat Toyah Basin leaving the Cretaceous escarpments and mesas behind. The basin continues to the northern point of the Davis Mountains just to the right of the freeway. The Barrilla Mountains are on the horizon from 9 to 10:30 o'clock.

The Highway 17 Exit to Pecos is at Mile 212.7. Continue on Interstate 10 to Mile 210 and take the Highway 17 exit to Fort Davis.

About 1½ miles from the exit, Highway 17 passes a bluff on the left of Huelster tuffs capped by a thin lava bed. The formation is named for the Huelster Ranch, now long abandoned, 4 miles south of Toyahvale. There, the formation is 400 feet thick but thins to the east. Although it outcrops widely in the Barrilla Mountains, the only easily accessible exposure is on the road to Balmorhea Lake from Balmorhea, where 2 miles south of town, light gray Huelster tuff is exposed in a quarry on the left of the road.

## Balmorhea

Balmorhea is a mile ahead, conspicuous by its trees, an oasis in the desert. The oasis effect does not come from rainfall, which averages only 13.59 inches here according to the National Weather Service, but rather from springs. Several springs are found along the valley between the mountains on the left and the hills on the right, the most important of which are the San Solomon Springs at the Balmorhea State Park, 4 miles west of Balmorhea. The valley is drained by Toyah Creek which runs just north of Balmorhea and then turns northeast to join the Pecos River near Pecos.

The town of Balmorhea was laid out in 1906 in the center of a 14,000-acre tract watered by San Solomon Springs and named for the three developers, Balcum, Moore, and Rhea. In 1911, the Pecos Valley Southern rail line was completed from Pecos and a hotel built. The population reached a maximum of 1,200 in the 1930s. The town declined as irrigation declined; the 2000 census gives Balmorhea a population of 527.

Midway through the town of Balmorhea, Farm Road 2903 on the right crosses Interstate 10 and leads to Toyah on Interstate 20, 23 miles away to the north. Across Interstate 10, the Meier Hills on the horizon are Pleistocene alluvial deposits of sand and boulder-laden gravel with scattered small outcrops of Gomez Tuff breaking through. The hills are 300 feet above Balmorhea, the depth of the Toyah Creek valley today.

Two blocks ahead, the sign to Balmorea Lake connects to a county road which leads to the lake. Gomez and Huelster tuffs can be seen in a quarry two miles from Highway 17. Continuing on Highway 17, Carpenter Hill (3,656 feet) ahead on the left is at the northern end of the Barrilla Mountains, an outlier of Star Mountain rhyolite overlying Huelster tuffs. Saddleback Mountain is seven miles away at 9 o'clock on the horizon.

Cement-lined water ditches follow the highway west from Balmorhea. They are part of the supplemental irrigation water supply to 10,608 acres within Reeves County Water Improvement District No.1. The district's primary sources of water are the San Solomon Springs and the smaller Giffen, Saragosa, and West Sandia Springs.

Fig. 3.3: Balmorhea State Park, a tranquil setting. The small bridge in the center foreground takes you over the canal carrying water from San Solomon Springs to the Lake Balmorhea reservoir.

Fig. 3.4: The beautiful symmetrical Star Mountain from the junction of Highway 17 and 3078. It has a thin cap of Barrel Springs welded tuff at its summit, underlain by Frazier Canyon tuffs, a thin bed of Gomez Tuff and 500 feet of Star Mountain Formation lava flows. The line of light near the summit is at the top of the lava flows. Compare to the cross-section (Fig. 4.3).

Fig. 3.5: Just west of Star Mountain, the escarpment continues with Forbidden Mountain, capped by a thick level flow of Adobe Canyon Formation rhyolite. The light-colored pyramid in front is Little Aguja Mountain. See also Fig. 6.6.

Fig. 3.6: The northern end of the escarpment at nightfall, from the junction of Highways 17 and 3078. Gomez Peak is on the right.

Irrigation in the Madera Valley dates back to 1870, when vegetables and feed crops were produced for troops and livestock at Fort Davis. After 1880, irrigation expanded rapidly and reached its peak in 1909 when the Toyah Valley Irrigation Company was organized by consolidating several small canal systems. It was acquired by the U.S. Bureau of Reclamation which constructed additional facilities in 1946-7. Diversion dams were built on Madera and Toyah Creeks and concrete canals were constructed to channel water from the dams to the State Park swimming pool, and from the pool to Balmorhea Lake, a reservoir south of Balmorhea, where it is held for irrigation.

An interesting study on the sources of water in the Balmorhea area springs was published by Chowdury, Ridgeway and Mace of the Texas Water Development Board in 2004. Using chemical analyses, they determined that although some of the spring water came from the Davis Mountains, about 80 per cent of it came from the Salt Basin north of Van Horn through fault channels in the Apache Mountains. Isotope analysis suggested that it had accumulated during humid periods in the Pleistocene era, 10,000 to 16,000 years ago. In effect, these springs are mining water from the Salt Basin.

## Balmorhea State Park

Balmorhea State Park is about 4 miles west of Balmorhea. The 46-acre area was developed by the Civilian Conservation Corps in the period 1935-40, one of many projects funded by the federal government to alleviate unemployment during the Great Depression, including Indian Lodge near Fort Davis and Chisos Mountains Lodge in Big Bend National Park. The CCC structures in the park, built in a Spanish Colonial style with stucco exteriors and tile roofs, include a limestone concession building, two bathhouses, an adobe superintendent's residence, and the San Solomon Courts, an early version of the modern-day motel, constructed of adobe bricks. The recreation area includes bungalows, campgrounds and a 1¾-acre artesian spring pool set among trees on a grassy flat.

The swimming pool is one of the largest man-made pools in the United States, containing 3.5 million gallons of clear, 76° Fahrenheit water fed by San Solomon Springs. Water flow from the springs averages about 20 million gallons per day, although in wet 2004 it rose to 35 million gallons per day. The pool has a large main circular area 25-foot deep with two long rectangular arms extending out. One arm is three to five feet deep; the other is 20 feet deep.

Scuba divers can achieve open water certification in the swimming pool because of its depth. The park has about 200,000 visitors a year from Texas, New Mexico, Colorado and Oklahoma.

The park has mature cottonwood trees, netleaf hackberry and Arizona ash, and native plants like catclaw mimosa, mesquite, creosotebush and whitebrush. The Civilian Conservation Corps planted fruiting mulberry, plateau liveoak and ornamental juniper in the 1930s which provide additional shelter and feeding habitat for birds. During migration the trees fill with brightly colored orioles, vireos and buntings. As many as 150 species can be seen at the park in the springtime and almost as many in winter.

A cienega, or wetlands area of canal and reeds was constructed in 1975 to provide a stable habitat for two endangered species of fish, the Comanche Springs Pupfish *(Cyprinodon elegans)* and the Pecos Gambusia *(Gambusia nobilis)*. You can look down on the fish from a walkway above the canal or look into the water through an inspection window. Five other kinds of fish can be seen in the canal; Largespring Gambusia, Rio Grande Tetra, Roundnose Minnow, Headwater Catfish and Green Sunfish.

At the scattered buildings of Toyahvale just beyond the park, Highway 17 turns south towards Fort Davis at the junction with Farm Road 3078. The journey from Toyahvale to Fort Davis is described in Chapter 4.

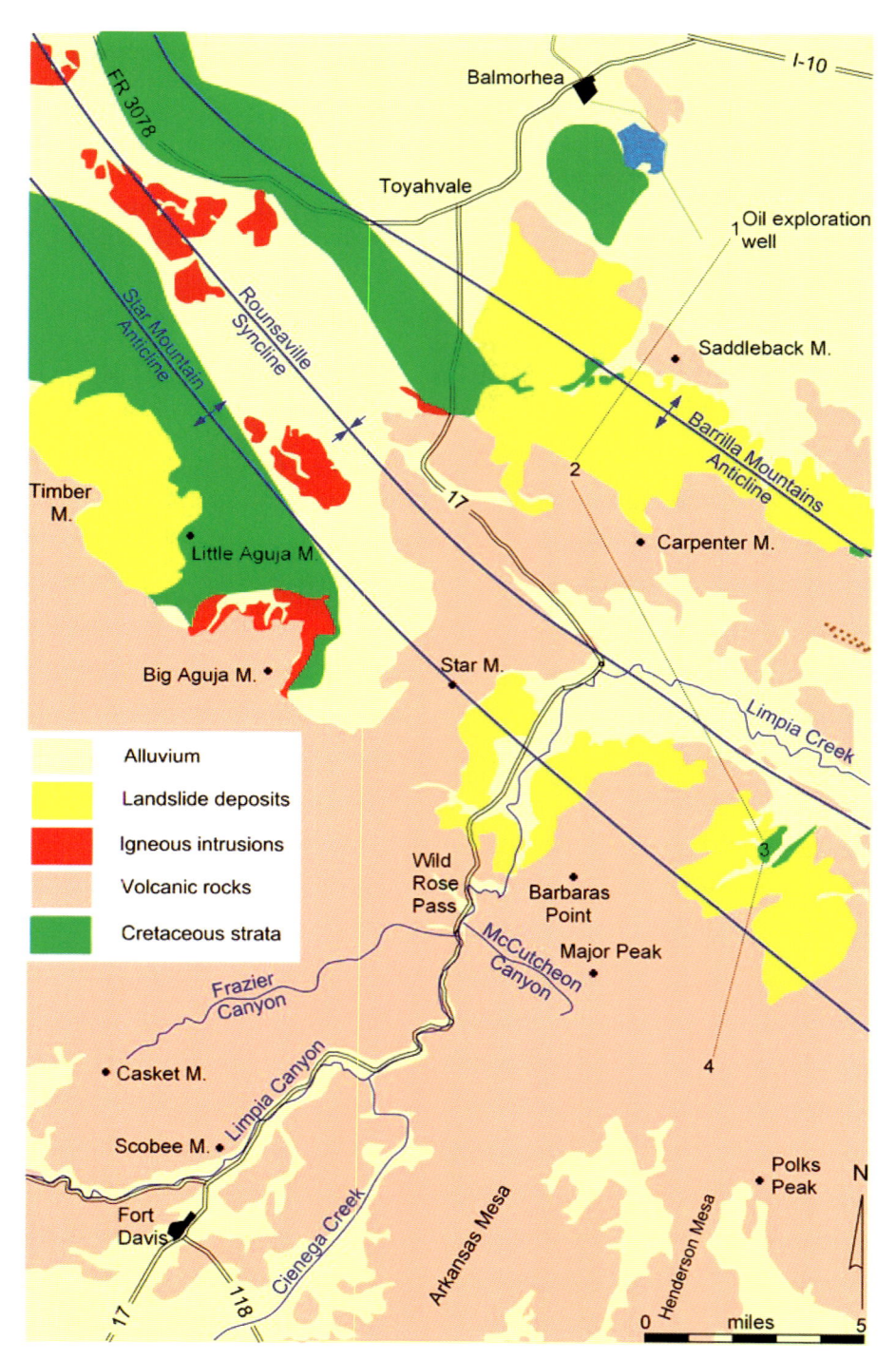

Fig. 4.1: Geology from Toyahvale to Fort Davis

# 4: TOYAHVALE – FORT DAVIS

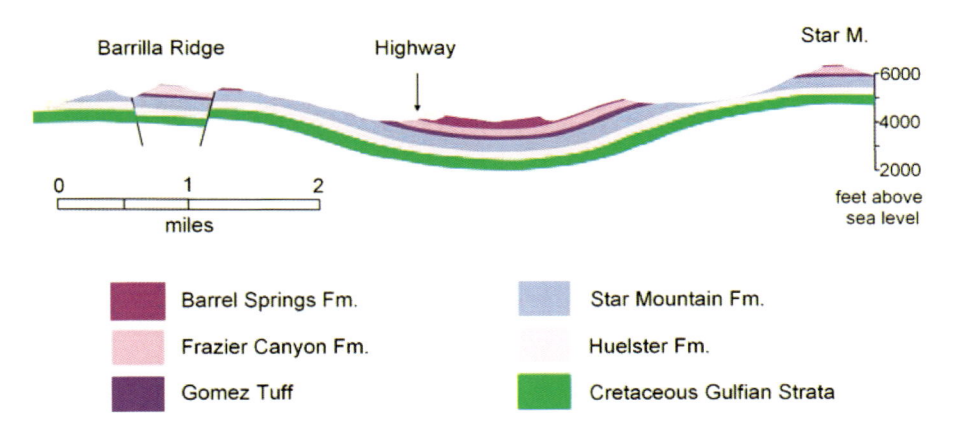

Fig. 4.2: This section across the Rounsaville Syncline from Star Mountain to the Barrilla Mountains ridge crosses the highway at Mile 8.9, 1.3 miles beyond Seven Springs Ranch. The apex of the ridge is a shallow graben, capped by a thin Barrel Springs bed. (Adapted from Eifler).

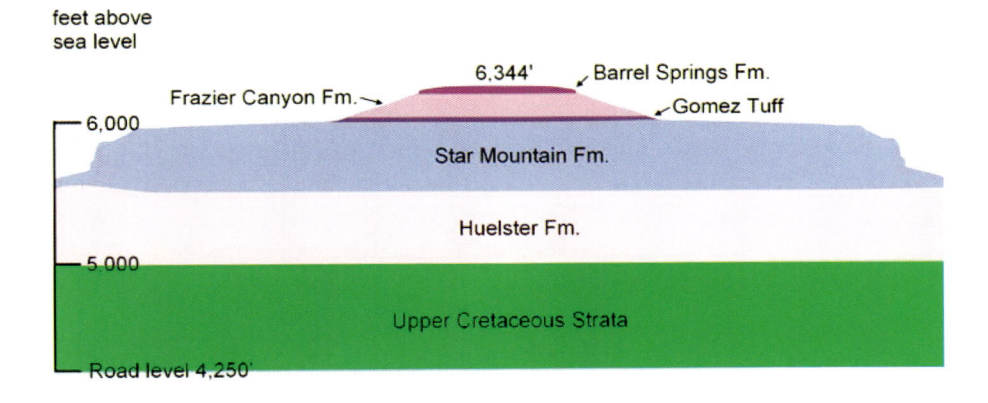

Fig. 4.3: Generalized section across Star Mountain from Mile 14.0 near Wild Rose Pass. The Star Mountain and the Huelster Formations are each about 500 feet thick, and the Frazier Canyon Formation about 250 feet thick. The upper Barrel Springs Formation and the Gomez Tuff are both around 50 feet thick. The Upper Cretaceous section thickness is not known nearby oil exploration wells intersected about 800 feet of the unit (see page 52).

# 4: Toyahvale – Fort Davis

## 32 Miles

Highway 17 turns south towards Fort Davis at the junction with Farm Road 3078, a quarter-mile west of Balmorhea State Park. Mileages for this segment are measured from the junction.

The panorama from the junction is quite exceptional, one of the best in the mountains. It extends from the low hills at 10 o'clock to Gomez Peak (6,320 feet) at 4 o'clock (the geological map, Fig. 4.1, continues to the west on Fig. 6.1). The escarpment from Star Mountain (6,344 feet) at 12 o'clock to Gomez Peak is the northern perimeter of a lava plateau of about 6,500 feet elevation, dissected by canyons up to 1,500 feet deep. The plateau is about 3,200 feet above the highway, a relief rarely seen in Texas.

Although volcanic rocks extended beyond the junction at one time, they have now been eroded away and the road begins on alluvium overlying Cretaceous Boquillas strata (the Cretaceous succession in this area is given in Fig. 10.2). Five miles ahead, it crosses on to volcanic rocks and climbs up through them to the Frazier Canyon Formation at Fort Davis, 1,600 feet above Toyahvale.

The lower volcanic sequence is well displayed on Star Mountain with, from bottom up, Cretaceous shale and limestone followed by Huelster bedded tuffs, Star Mountain quartz trachyte lava, 500 feet thick, which forms the bold lower cliff, a thin bed of Gomez ash-flow tuff at the top of the cliff, Frazier Canyon volcaniclastic strata and Barrel Springs ash-flow tuff at top (Fig. 4.3).

In Timber Mountain at 2 o'clock, the sequence is, from bottom up, Huelster, Star Mountain and Gomez Formations, overlain by Adobe Canyon rhyolite, a prominent lava in the northwest mountains. Frazier Canyon and Barrel Springs strata have been eroded off Timber Mountain. A quarter mile from the junction, the highway crosses the Madera diversion canal, which carries water from a diversion dam in Madera Canyon to the swimming pool at Balmorhea State Park.

# 4: TOYAHVALE – FORT DAVIS

| Unit | Age m.y. | Description |
|---|---|---|
| Barrel Springs Formation | 35.6 | Three units seen between Toyahvale and Fort Davis, a lower welded ash-flow tuff, a middle bedded sedimentary unit 160 ft. thick and an upper welded tuff or lava. |
| Sleeping Lion Formation | 35.9 | Single rhyolite flow in central & southeast Davis Mtns., probably from source northwest of Fort Davis; 630 ft. thick at maximum. |
| Frazier Canyon Formation | undated | Tuff and tuffaceous sandstone with lenses of pebble conglomerate; Frazier Canyon reaches maximum thickness of 1,150 ft. in eastern Davis Mountains. |
| Adobe Canyon Formation | 36.5 | Two or possibly three rhyolite flows, each 400-850 ft. thick, in north central & northwestern Davis Mountains. |
| Limpia Formation | | Quartz trachyte lavas overlying Gomez Tuff east & southeast of Fort Davis. |
| Gomez Tuff | 36.7 | Rhyolite ash-flow tuff in northern & eastern Davis Mountains; source is Buckhorn Caldera; up to 2,000 ft. thick in caldera, abruptly thinning to a maximum of 480 ft. outside the caldera. |
| Star Mountain Formation | 36.8 | Multiple rhyolite to quartz trachyte lava flows in eastern Davis Mountains; individual flows 200-600 ft. thick. |
| Huelster Formation | 38.4 | Reworked tuffs; mafic lavas near base of Huelster Formation in N.E. Davis Mountains |
| Upper Cretaceous undivided | 98-45 | Marl, shale and clayey limestone; 300 ft. thick. |
| Upper Cretaceous Boquillas | | Upper part, interbedded marl and shale; lower part, limestone, silty to sandy, flaggy, dark grayish orange near base; marine megafossils; 200 ft. thick. |

Fig. 4.4: Formations seen along Highway 17 between Toyahvale and Fort Davis.

The low hills on the left at the Reeves-Jeff Davis county line (Mile 2.8) are small remnants of Star Mountain rhyolite above Huelster tuff with landslide deposits on their flanks. The distinctive profile of Saddleback Mountain (4,286 feet) is at 9 o'clock on the skyline (Fig. 4.5). The upper ledge is formed by the lowest Barrel Springs member, a welded ash-flow tuff, which is at roadside 7 miles ahead, and also caps Star Mountain. These elevation differences are partly due to folding (discussed below) and faulting. A fault between the Barrilla Mountains ridge and Saddleback has displaced strata down 700 feet. Strata on the mountain rise towards the Barrilla Mountains anticline, creating the saddleback profile. Along this straight, Big Aguja Mountain is at 1 o'clock, the flat topped Timber Mountain at 2 o'clock and behind it, Forbidden Mountain, also flat topped.

Fig. 4.5: Saddleback Mountain from Mile 1.0. The upper ledge is formed by the lower Barrel Springs welded ash-flow tuff. Strata rise towards the Barrilla Mountains anticline on the right.

Fig. 4.6: The rugged northern end of the Barilla Mountains ridge, photographed here from Mile 4.5. These peaks are capped by Star Mountain lava underlain by Huelster tuffs with landslide blocks of the lava and tuffs at lower elevations.

Fig. 4.7: Forbidden Mountain on the skyline from Mile 5.8. The cliff at top is a thick flow of Adobe Canyon rhyolite. Little Aguja Mountain is in front at right, an intrusive with a Star Mountain cap. The sharp peak to the left of Little Aguja Mountain has an intrusive core that crops out at its summit, with upper Cretaceous marls beneath which appear as white patches near its base. The lower Barrel Springs welded tuff outcrops in the foreground are near the Rounsaville Syncline axis which runs across the photograph.

Fig. 4.8: Big Aguja Mountain from Mile 5.8. The lower smooth slopes are underlain by a quartz trachyte laccolith intruded between Cretaceous and Star Mountain strata, the latter cropping out in the upper section of the mountain.

At Mile 3.4, yellowish gray weathered Upper Cretaceous shale crops out on a ridge on the right of the road and again in the face of an old road quarry. A trachytic intrusion forms the hill on the right at Mile 3.9. Small outcrops and rubble from the intrusion are at roadside on the right.

## Stop 1: The Rounsaville Syncline (6.3 miles from Toyahvale)

The highway turns sharply left into the Rounsaville syncline bringing into view one of the most stunning vistas in the Davis Mountains. The road follows the syncline, a flat-bottomed, steep-sided cleft in the mountains for the next 6 miles. On the left, a ridge in the Barrilla Mountains rises up to 1,500 above the road, on the right a ridge is from 500 to 750 feet above the road.

The cleft, although it has ephemeral streams running north to join Aguja Canyon on the right, is not a stream or river valley. Rather it developed from its underpinnings collapsing below it. This deep, narrow fold is unique in the volcanic Trans-Pecos landscape and geologists have struggled to account for it. Henry, Price, Parker and Wolff make the comment that the syncline "probably resulted from a sag into a minor Basin and Range graben."

Bruce Pearson in a 1985 paper drew a section through the four oil exploration wells in Fig. 4.1 showing that the volcanic rocks are underlain by 800 feet of soft Upper Cretaceous marl. Deeper down, he showed that the axis, the lowest point, of the syncline nearly coincides with the edge of the Delaware Basin where limestones change to evaporites (Figs. 10.3 and 2.6). The soft marl and the equally soft evaporites, mainly salt and gypsum, are incompetent, that is, they flow rather than fracture under pressure. Pearson concluded that flowage, movement and solution of the evaporites may have contributed to the deep sag in the syncline.

Big Aguja Mountain (5,722 feet) is directly ahead as you make the turn. The mountain is underlain by a quartz trachyte laccolith intruded between Cretaceous and Star Mountain strata. Parker and Gilmore, based on chemical analysis, propose that the laccolith was a source for several of the Star Mountain lava flows. Its upper section is of Star Mountain rhyolite overlain by a thin section of Gomez Tuff. Several lava beds crop out on its flanks and on its summit.

The butte-like Little Aguja Mountain (5,192 feet) is at 1 o'clock. The light-colored rocks on its flanks are Upper Cretaceous shales and limestones. Above them a rhyolite intrusion crops out on the smooth light brown slopes, perhaps a continuation of the chemically-similar Big Aguja Mountain laccolith.

The Buffalo Trail Scout Ranch entrance is on the right at Mile 6.6. The 9,500-acre ranch is used for 8 weeks every summer by 13,000 scouts from 13 West Texas counties, including Jeff Davis, Brewster and Presidio.

A half-mile ahead, strata dipping off the Star Mountain anticline into the Rounsaville syncline can be seen beyond the Seven Springs ranch buildings. On the left of the highway, strata dip to the right in the other flank of the syncline. The trees at the ranch buildings grow around the springs for which the ranch is named and which are at the axis or lowest point of the syncline.

From the ranch entrance at Mile 7.6, the highway climbs up the syncline valley through the Barrel Springs Formation. The lowest unit of this formation, a welded ash-flow tuff, crops out around the ranch buildings and is exposed in a road cut on the right 1½ miles beyond the entrance. It is overlain by 160 feet of white bedded volcaniclastic strata which is exposed intermittently ahead and by an upper welded ash-flow tuff or possibly a lava which can be seen in a right road cut at Mile 10.9.

Along this stretch of road, the terrain slopes from left to right, signaling that the syncline axis is to the right of the highway. At Mile 11.0, cuestas on both sides of the road dip into the nearby syncline axis (Fig. 4.9). In a cuesta, one side of a ridge slopes down parallel to bedding in the underlying rock while the other side forms a steep escarpment. The terrain opens out at this point to an attractive broad flat.

## Stop 2: Barrilla Mountain Ranch Entrance (12.7 miles from Toyahvale)

The Rounsaville Syncline axis crosses the highway shortly before this entrance as the road turns hard right. Just beyond the entrance, Limpia Creek turns sharp left to follow the axis.

Although Lieutenant Whiting did not identify his route from Fort Stockton in 1849, the party probably came up the Limpia Creek valley because he does not mention the springs at Balmorhea, which he certainly would if he had come that way. At some point near here, his party met a large band of Apaches:

> "They sternly demanded who we were and whether we came to the Apache country for peace or war. I answered Americans, en route for Presidio. We came peaceable; if we remained so, depended on them. In the meantime the mules had been tied up, their heads together; and in front of them appeared the Texans, squatted to the ground, their rifles cocked, their mouths filled with bullets, and their faces showing every variety of determined expression from the angry flush on the face of the younger men to the cool indifference of the veteran of San Jacinto and the men of Mier and Monclova. They waited but for a signal."

Fig. 4.9: The upper Barrel Springs welded tuff or lava dips off Star Mountain into the Rounsaville Syncline at Mile 11.8. The summit of the mountain is on the horizon at left. Although they do not show in the photograph, the white middle tuff crops out in places under the welded tuff.

Fig. 4.10: The slope at Mile 13.4 is capped by the Barrel Springs lower welded tuff on the extreme right underlain by Frazier Canyon Formation in which the gray outcrop is basalt. Below are rubble and outcrops of red-brown Gomez Tuff followed by a gap and finally Star Mountain lava at roadside just to the left of the white windmill tank. The Frazier Canyon Formation is quite thin here but thickens considerably towards Fort Davis.

Fig. 4.11: Wild Rose Pass from the flat at Mile 15.8. The pass that Whiting named is to the left of the pyramid on the skyline, where the creek runs along the base of the dark cliffs. At least two Star Mountain units appear on the cliffs to the left of the pass with a break in the cliffs between them. Today's pass is at the base of the cliffs on the right. The rough ground between the two is capped by Star Mountain lava overlying Huelster tuff.

Fig. 4.12: Star Mountain from the flat at Mile 16.6. Two flows of Star Mountain rhyolite are separated by a layer of light gray breccia in an ancient or paleo-valley. The thin Barrel Springs bed at the summit and the thin Gomez Tuff bed just above the cliffs stand out clearly. Huelster tuff underlies the Star Mountain lavas and outcrops in places (see cross-section at Fig. 4.3).

He was able to persuade the Apaches to confer and the party was led to their encampment:

> "Cautiously and with much apprehension on the part of my men of treachery from the savages, we followed the yelling bands. It was an exciting and picturesque scene. Two hundred Apache, superbly mounted, set off by their many colored dresses, their painted shields, and hideous faces, galloping to and fro and full speed and brandishing their long lances, moved in advance down the valley. Behind at some distance, mounted humbly on jaded mules, close together and watchful, came the little band of Texans."

After an evening of discussion and visiting the Apache township the following day, the party led off south through Wild Rose Pass without event except for the purloining of many small articles, including all Whiting's papers, the latter much sought after for making cigarillos.

The highway follows Limpia Creek upstream for the next 18 miles to the junction with Highway 118. A cuesta on the right bank of the creek dips to the left, showing that the syncline axis is behind us. Just before the road again bends to the right at Mile 13.3, Barrel Springs lower welded tuff crops out at the top of the slope on the right, followed by a section of Frazier Canyon Formation in which the only outcrop is of gray basalt. Next are dark red-brown outcrops and rubble of Gomez Tuff followed by another gap and then at road level, the Star Mountain lava (Fig. 4.9).

At Mile 14.0, the highway is on a flat between high palisaded cliffs on either side. Star Mountain on the right rises 2,100 feet above the road, its upper reaches creating an elegant curved profile. The small subsidiary dome on top is capped by the lower Barrel Springs ash-flow tuff which has protected the underlying soft Frazier Canyon tuffs from erosion (Fig. 4.12).

At Mile 15.6, at least two Star Mountain flows are visible in the wall of Limpia Canyon at 11 o'clock, with a break in the escarpment between them. At the top of the cliff, a thin cap of Gomez Tuff is separated from the Star Mountain strata by yellow, bedded tuff 10 or 12 feet thick. On both sides of the highway, landslides of Huelster tuffs and Upper Cretaceous mudstones and claystones create hummocky terrain at the bottom of the cliffs.

From the ranch entrance at Mile 16.6 the high ground at 9 o'clock is Barbaras Point (5,992 feet), an outlier of Barrel Springs over Gomez Tuff and Frazier Canyon strata. On the right, Star Mountain is about a mile back from the road with two flows of Star Mountain Formation, each 300 feet thick or more, separated by light-colored breccia which fills an ancient valley between them. Small outcrops of Huelster tuff can be seen below the cliffs (Fig. 4.10).

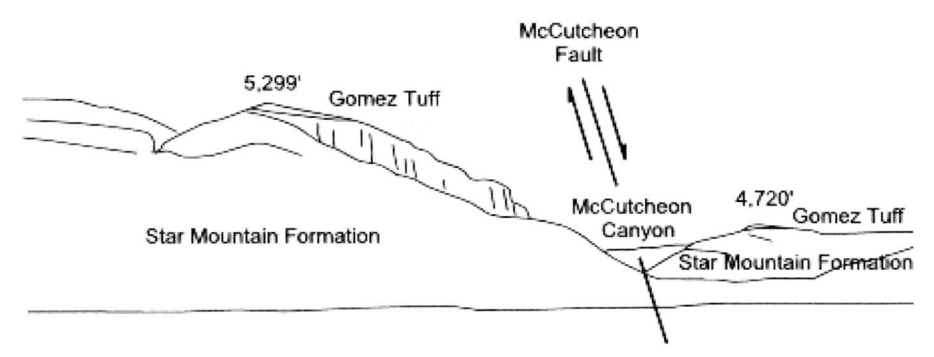

Fig. 4.13: The McCutcheon Fault looking left at Mile 19.9. The bluff above the fault is capped by Gomez Tuff as is the rise on the right of the canyon, giving a displacement of about 570 feet down to the right.

Upper Cretaceous mudstone is exposed in a road cut on the right at Mile 17.2. Oyster and clam fossils as well as gypsum crystals can be found in the weathered zone.

## Stop 3: Wild Rose Pass (18.3 miles from Toyahvale)

The road begins the climb up a notch in the ridge ahead at Mile 18.3. Such gaps are called wind gaps by geologists (Fig. 4.11). They are generally higher than water gaps. The real Wild Rose Pass is a water gap about a mile to the east where Limpia Creek has cut a canyon through a jumble of landslide blocks of lava and alluvium.

You can see the canyon entrance on the left at about 10:30 o'clock, the original pass named by Whiting:

> "We set out from camp early. Our road has been remarkable. The daylight showed us, on awaking, a fine pass. Through the gorge, now running at the base of the dark cliffs of basaltic columns, now winding amid the prettier groupings of trees and in little mountain valleys, is a clear stream. We followed it through the range, delighted at the promise of a successful passage of the road where wood and water should obtain in plenty. Wild roses, the only ones I had seen in Texas, here grew luxuriantly. I named the defile the "Wild Rose Pass" and the brook the "Limpia"."

Although fine for horses and mules such as those used by Whiting's party, the pass is prone to flooding in the rainy season, and wagoners soon took to using the path now followed by the highway. The roses have been identified as Apache Plume, *fallugia paradoxa*, a member of the rose family. The creek rejoins the highway about a mile and a half ahead.

Fig. 4.14: Lava flows or welded ash-flow tuffs contract as they cool and develop vertical joints. Erosion along the joints creates lava columns, shown here in the Star Mountain Formation in Limpia Canyon at Mile 21.3. Geologists call such cliffs palisades. The name comes from The Palisades, cliffs along the Hudson River in New York and New Jersey.

Fig. 4.15: This photograph from Mile 23.5 shows the relationships between the Sleeping Lion Formation, 230 feet above the highway on the left, the Gomez Tuff at top right, and the Star Mountain Formation in front center. The slopes below the upper cliffs are underlain by tuffs and basalt flows of the Frazier Canyon Formation which crop out sporadically between here and Fort Davis.

Fig. 4.16: The northwestern end of a mesa capped by Sleeping Lion jointed lava in the morning sunlight at Mile 24.9. Gomez Tuff crops out among the trees on the left and on the ridge on right front.

Fig. 4.17: Gomez Tuff at roadside at Mile 25.3. Although the weathered tuff is usually a dark chocolate brown, fresh surfaces are pink.

Star Mountain Formation caps the ridges on either side of the pass underlain by the smooth yellow-brown slopes of a basalt sill intruded into Huelster Formation strata. In a road cut at the summit (Mile 18.7), round boulders of basalt are set in weathered material, an example of spheroidal weathering, in which jointed igneous rock gradually erode into sphere-like shapes as they erode.

Just beyond the summit of the pass, the creek swings back to the road between 800-foot high cliffs following a minor fault crossing the road. The fault presumably created a weak zone for the stream to exploit. The highway enters Limpia Canyon at Mile 19.3 and follows it almost to Fort Davis. Limpia Creek crosses the road repeatedly for the next several miles.

## Stop 5: McCutcheon Fault (20.1 miles from Toyahvale)

At the stock pens, McCutcheon Canyon enters from the left. A fault, named the McCutcheon fault by geologists, crosses the road just before the pens, dropping strata nearly 600 feet down to the right (Fig 4.13). The high cliffs on the left of the canyon are capped by Gomez Tuff, while on the right of the fault the tuff caps a cuesta that dips into the fault at 3 o'clock. The fault dies out to the northwest.

The highway crosses Frazier Canyon Creek and passes a picnic area at Mile 20.6. The Star Mountain Formation upper flow crops out around the picnic area and along Limpia Canyon for the next 5 miles in a series of beautiful palisaded cliffs (Fig. 4.14).

At another picnic area on the left at Mile 22.6, thin reddish-brown Gomez Tuff caps Star Mountain Formation. Over the next 2½ miles, the road rises 150 feet to the level of the tuff which is exposed in a small road cut at Mile 25.3 (Fig. 4.17). The high ridge in the background at 12 o'clock has Barrel Springs overlying Sleeping Lion and Frazier Canyon Formations (Fig. 4.16).

Limpia Formation makes its first appearance in a road cut at Mile 27.3, in which its base is exposed, and is exposed in road cuts on the right for the next mile (Fig. 18). It overlies Gomez Tuff, from which it is typically separated by several feet of bedded tuffaceous sedimentary rock, and is overlain by the Sleeping Lion Formation in Limpia Canyon. It forms smooth, rounded slopes and creates boulders that were used by early settlers to build stone fences, a beautiful example of which runs along the roadside for the next half-mile (Fig. 4.19).

The historical marker at Mile 30.1 describes the work of Barry Scobee, a major figure in the rescue of the old Fort Davis from decay. Scobee Mountain (5,420 feet), on the right was named for him.

At Mile 30.3, the old fort is visible at 12:00, nestled in Hospital Canyon. At 1:00, Blue Mountain is on the horizon up Limpia Canyon. The junction of highways 118 and 17 is at Mile 30.7. The ridge paralleling the road to the right has Frazier Canyon Formation at base, overlain by cliff-forming Sleeping Lion rhyolite, capped by Barrel Springs ash-flow tuff.

After leaving Wild Rose Pass, the Whiting party camped for their noon meal at the cottonwoods along Limpia Creek on the right. They had intended to camp for the night but, alarmingly, Apache smoke signals to the north and south indicated that an attack was being planned, so they pretended to set up camp and at 8 o'clock, after nightfall,

> "Our fires still burning, we issued from our camp without a sound. No man spoke. I may live a long time yet, but I shall never forget the still and oppressive hours of that somber march. The night was clear in its cold starlight. The wind swept by over the bleak plain in fitful and furious gusts from the west; to the eastward above the hills of the pass rose the lurid glare of gathering fires of the Apache. Anxious and with senses keenly alive to every sound, we moved in close order over the plain, listening each moment for the Indian whistle and the rushing of his horse's hoofs. We had left the trail as an additional precaution and were steering a west course.
>
> About eleven, on our left, a brilliant fire suddenly flashed into light from the summit of a lofty peak. The whispered words "a signal", "a signal" ran through our party. We knew that our movement had been discovered. The lurking spies about our camp had found we had gone and made all speed to their signal hill. How intense the excitement of the next two hours! That fire flashing on our left; while every little space the dark figures distinctly visible, would pass between us and the light, heaping on the brush and feeding the flame. Answering back in the east were the mountains, lit up with a long red line of fire."

By not taking the Presidio road down Alamito Creek, but rather going west towards Blue Mountain, they had avoided the ambush that had been set for them. They camped about 11 o'clock that night and safely made their way the next day down to Cibolo Creek and thence to Leaton's fort on the Rio Grande, now the site of the Fort Leaton State Park.

The Sleeping Lion Formation, a gray porphyritic rhyolite, forms the prominent cliffs framing Hospital Canyon behind the Fort Davis National Historic Site (Mile 31.1), 200 feet thick on the ridge on the west of the canyon. The formation overlies the Frazier Canyon Formation here; a basalt flow in the Frazier Canyon Formation crops out in the creek between the parking lot and the visitor center.

The Limpia Hotel at Mile 31.8 and the Fort Davis Bank building were built partly with pink welded ash-flow tuff from the base of the Barrel Springs Formation. The Jeff Davis County courthouse is at Mile 31.8.

Fig. 4.18: Limpia Formation from Mile 27.6. The rock is fractured vertically and horizontally and breaks easily into slabs.

Fig. 4.19: A stone fence built from Limpia Formation boulders at Mile 27.7.

Fig. 4.20: Looking up Hospital Canyon from the parade ground at the Fort Davis National Historical Site. Sleeping Lion Mountain is on the left, capped by Sleeping Lion Formation rhyolite. The ridge on the right, also capped by Sleeping Lion lava extends up Limpia Canyon to the Davis Mountains State Park. The buildings in front housed officers.

Fig. 4.21: Another view of the ridge to the right of Hospital Canyon. The four two-storey buildings below the cliffs were also quarters for the officers, built shortly before the fort closed in 1884. The white gable wall is of the chapel.

## Fort Davis

Lieutenant Whiting returned to San Antonio in May, 1849, 3½ months after his departure, and word spread very quickly about his discovery of a useable route to the west. General Harney, now the commander of the Eighth Military District in succession to General Worth, who had died of cholera, ordered Colonel Joseph Johnson, the army's chief topographical engineer, to organize excursions to examine the Davis Mountains route and a northern route along the Guadalupe Mountains.

The Davis Mountains party included Johnson and Colonel Jefferson Van Horne who commanded a wagon train of 275 wagons and 2,500 horses and mules carrying the stores and weaponry of the Third Infantry Battalion. The battalion was to establish posts at what is now Fort Bliss at El Paso and up and down the river.

Johnson went ahead of the main wagon train with about 20 civilians who cleared the way for a small wagon train he took with him, thus creating the first route for wheeled vehicles through the Davis Mountains. They arrived in El Paso in September.

The California gold rush was on in full force. According to Roy Swift in *Three Roads to Chihuahua,* a current report estimated that in July, 1849 there were 1,200 wagons and 4,000 emigrants camped at El Paso, waiting for direction west. Commercial traffic from the Gulf of Mexico to Santa Fe quickly developed, too, as well as traffic to supply the army at Fort Bliss.

Raids on the wagon trains were commonplace and in 1854, as part of its broader strategy to defend the routes west from San Antonio, the army decided to establish a fort at Painted Comanche Camp. It was named after Jefferson Davis, the Secretary for War. The first fort had officers' quarters and a hospital built of wood scattered up Hospital Canyon and a line of stone barracks at the canyon's mouth. It was later rebuilt on the present site in 1867, after the Civil War ended.

A settlement grew up near the fort, initially down Limpia Creek where water was available, as pioneers were drawn to supply the fort and by the safety of living under the Army's protection. When Presidio County was organized in 1871, Fort Davis was selected as the county seat. The townspeople rejected a proposal by the Galveston, Harrisburg and San Antonio Railway to route its railroad through the town in 1882, and when the railroad instead went through Paisano Pass to Marfa, certain citizens, led by an attorney, John Dean, who had bought up land in Marfa, were able to force another election in 1885, in which Marfa was chosen as the county seat. The losers agitated to split off from Presidio County and were able to have Jeff Davis County established in 1887, with Fort Davis as the county seat.

The population of the town, which was estimated to be around 2,000 in 1885, declined after the fort closed in 1891 to about 500 in 1896 but has fluctuated between 700 and 1,200 since then. In the 2000 census, it was reported to be 1,050. The town today has a flourishing tourist industry. Texans are drawn to the mountains in the summer, when Fort Davis is the coolest place in the state on the Fourth of July. In the fall, northerners come through the town on their way to the Rio Grande valley and again in the spring on their return journey.

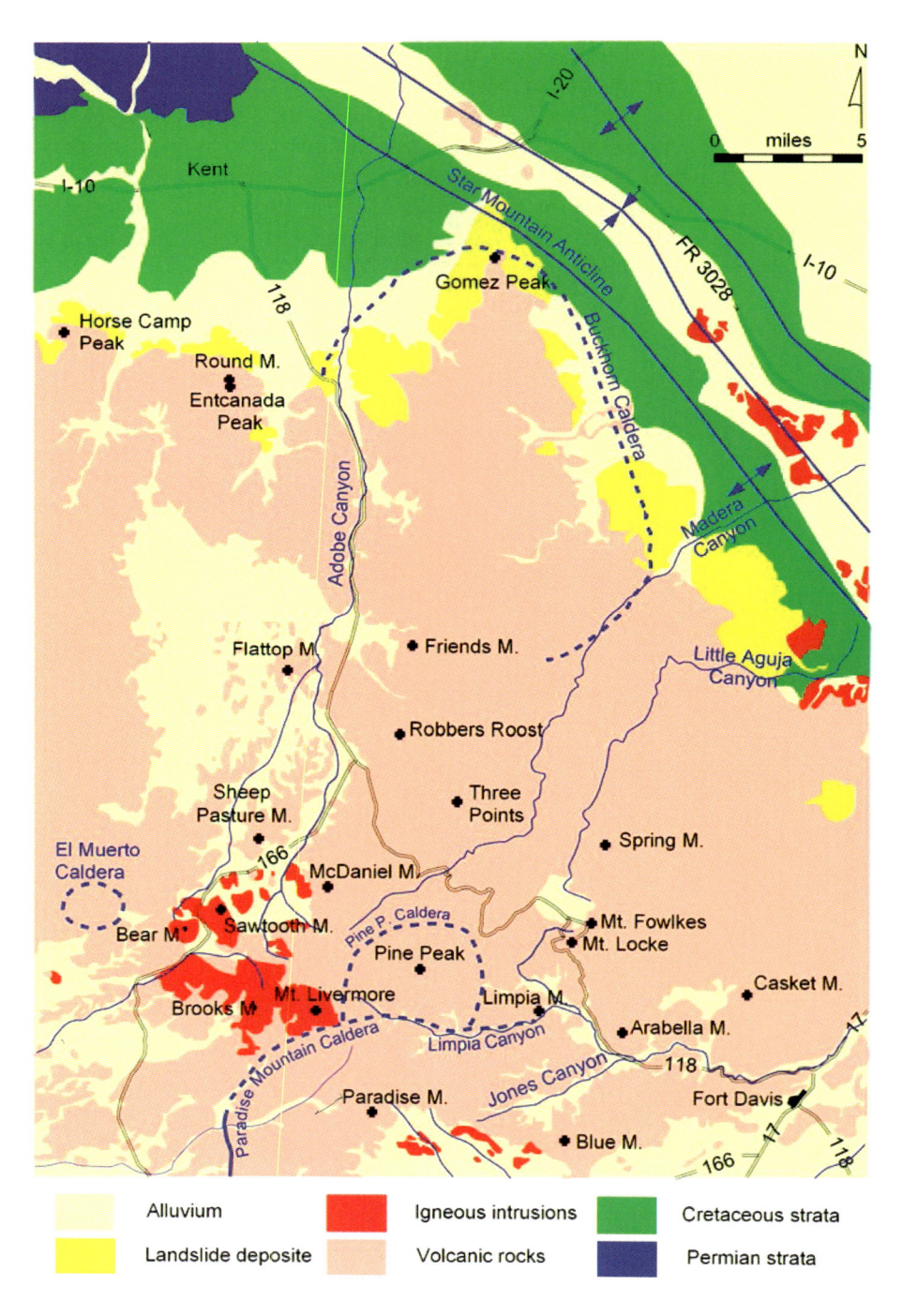

Fig. 5.1: Geology along the Fort Davis-Kent highway.

# 5: Fort Davis – Kent

## 51 miles

The Kent road takes you through the heart of the mountains, through the Davis Mountains State Park and the Nature Conservancy's Davis Mountains Preserve to the junction with the Scenic Loop, 29 miles north of Fort Davis. Then, in complete contrast, you cross an alluvial plain and descend through Adobe Canyon to the limestone Cretaceous flatlands around Kent. Along the way, the road passes by the McDonald Observatory, with its telescopes on Mounts Locke and Fowlkes at almost 6,800 feet, the highest point on Texas highways.

Set your odometers at the junction of Highways 17 and 118, about one mile north of the Jeff Davis County courthouse. The geology along the segment is given in Fig. 5.1. The formations you will see along the route are summarized in Fig. 5.2.

Highway 118 bears left where thick cottonwood groves line Limpia Creek. Painted Trees Road, just beyond the junction, commemorates the name given to the area by the Lieutenant Whiting party on their return journey to San Antonio. They had gone up river to what is now El Paso and after resupplying came back through the Davis Mountains and camped here. The trees which, on their outward journey, had crude drawings made by Comanches, now had drawings by Apaches mocking the Americans for running away (see page 61).

For the first 2½ miles, the highway winds along Limpia Creek among tuffs and basalt flows of Frazier Canyon Formation at road level on the left (Fig. 5.4) with Sleeping Lion rhyolite cliffs high above. Sleeping Lion boulders have come down to roadside in places (Fig. 5.3).

The Limpia Canyon Primitive Area entrance, the newest addition to the Davis Mountains State Park, is on the right at Mile 2.6. The 1,350-acre area includes the high ground across the canyon on the right. It was purchased by Texas Parks and Wildlife Department (TPWD) in 1990 and has 5 miles of backcountry hiking trails and 6 primitive camping sites.

| Unit | Age m.y. | Description |
|---|---|---|
| Mount Livermore volcanic dome | undated | Trachyte volcanic dome; up to 1,400 feet thick on summit of Mount Livermore. |
| Wild Cherry Formation | 35.3 | Fine-grained rhyolite tuff, up to 355 ft. thick, produced from Paradise M. Caldera in eastern Davis Mountains. |
| Casket Mountain lavas | | Up to 5 flows of porphyritic rhyolite capping ridges in southern, central and eastern Davis Mountains; 1,600 ft. thick on Blue M.; probably erupted from widely spaced fissures. |
| Mount Locke Formation | undated | Quartz trachyte & rhyolitic porphyry in central Davis Mountains, gray when fresh to brownish gray to reddish brown when weathered; maximum thickness 580 ft.; source unknown. |
| Barrel Springs Formation | 35.6 | Upper unit: rhyolitic lava. Lower unit: strongly rheomorphic ash-flow tuff or lava in central and eastern Davis Mountains; source presumably buried in central Mountains; up to 500 ft. thick. |
| Sleeping Lion Formation | 35.9 | Single rhyolite flow in central & southeast Davis Mtns., probably from source northwest of Fort Davis; 630 ft. thick at maximum. |
| Frazier Canyon Formation | undated | Tuff and tuffaceous sandstone with lenses of pebble conglomerate; Frazier Canyon reaches maximum thickness of 1,150 ft. in eastern Davis Mountains; source or sources unknown. |
| Adobe Canyon Formation | 36.5 | Two or possibly three rhyolite flows, each 400-850 ft. thick, in north central & northwestern Davis Mountains; source or sources probably buried in northern Davis Mountains. |
| Gomez Tuff | 36.7 | Rhyolite ash-flow tuff in northern & eastern Davis Mountains; source is Buckhorn Caldera; up to 2,000 ft. thick in caldera, abruptly thinning to a maximum of 480 ft. outside the caldera. |
| Huelster Formation | 38.4 | Reworked tuffs with mafic lavas near base in N.E. Davis Mountains; source unknown. |

Fig. 5.2: Formations seen along Highway 118 between Fort Davis and Kent.

## Stop 1: Davis Mountains State Park (2.7 miles from Highway 17)

The entrance to the park is on the left at the top of a rise. The 2,700-acre park is worth visiting for the sheer beauty of the setting and especially for the scenic Skyline Drive. The drive takes you up and along a ridge to a fine viewpoint at its east end where you can look down on Fort Davis and the countryside to its south and east.

The 1,350 acres of the park on this side of the highway were deeded to the Texas Parks and Wildlife Department in 1933 by the local Merrill family, and in the same year, the Civilian Conservation Corps began construction of Indian Lodge and several other buildings in the park. The park was opened to the public in late 1930s. The CCC buildings are all designed in southwestern style, very attractive in their own right, although not much in keeping with local architecture. Indian Lodge is being renovated by TPWD as this book is being written in the summer of 2005.

The main park road runs from the entrance along Keesey Canyon to Indian Lodge between ridges 500 to 600 feet high capped by Barrel Springs ash-flow tuff and lavas. Campsites and camping facilities are scattered below the road to the left along the mostly dry Keesey Creek.

## Stop 2: Skyline Drive

The paved Skyline Drive branches off the main park road on the left about a half-mile from the entrance, passes the Interpretive Center, and winds up a ridge. Going up the ridge it gives you a wonderful series of views up Keesey Canyon to Indian Lodge and Blue Mountain. The Barrel Springs Formation is nearly continuously exposed in road cuts providing you a unusual opportunity to examine its complex makeup.

The formation is described in detail by Henry, Price, Parker and Wolff. At least three tuff units are present separated by thin clay beds and all are well displayed in the climb up the ridge. The lowest unit, about 460 feet thick, begins in a road cut at the first hairpin corner. It has 13 feet of devitrified ash-flow tuff at base overlain by foliated rheomorphic tuff with half-inch partings. Going up the straight to the next hairpin bend, the foliation becomes more obvious and more closely spaced. Folding and minor brecciation appear near the hairpin with the foliation becoming more a flow banding.

After a short covered interval at the bend, the unit is intensely flow-folded and brecciated i.e. broken into angular fragments up to 6 feet in diameter (Fig. 5.5). Halfway along the next straight, a second unit, flow-banded strongly rheomorphic tuff or lava, overlies the irregular upper surface of the first, the two flows being separated by a 3-foot thick clay layer, perhaps the base of the upper unit. Around another corner in front, a middle flow crops out in a road cut on the left. This flow is not brecciated and contains many vesicles in its upper section, cavities that formed when gas bubbles were trapped in the material as it solidified.

The road then goes round a sharp hairpin bend where a driveway on the right leads to a viewpoint overlooking Keesey Canyon to Indian Lodge and Blue Mountain (Fig. 5.6). The main drive continues along the ridge for

Fig. 5.3: Lichen-covered boulders of Sleeping Lion rhyolite at Mile 2.0 along Limpia Canyon. Lichen is uncommon in Texas except in the cooler high altitudes of the Davis and Chisos Mountains.

Fig. 5.4: Frazier Canyon Formation at Mile 2.3 in Limpia Canyon. A thin pinkish basalt flow crops out near the top of the road cut. Basalt is much less viscous than rhyolite and can be found in quite thin flows.

Fig. 5.5: The upper part of the second Barrel Springs unit on Skyline Drive. The rock is a flow breccia in which blocks of flow-banded lava are set in a devitrified and hydrated matrix.

Fig. 5.6: Looking up Keesey Canyon from Skyline Drive to Indian Lodge. Blue Mountain is on the horizon at left.

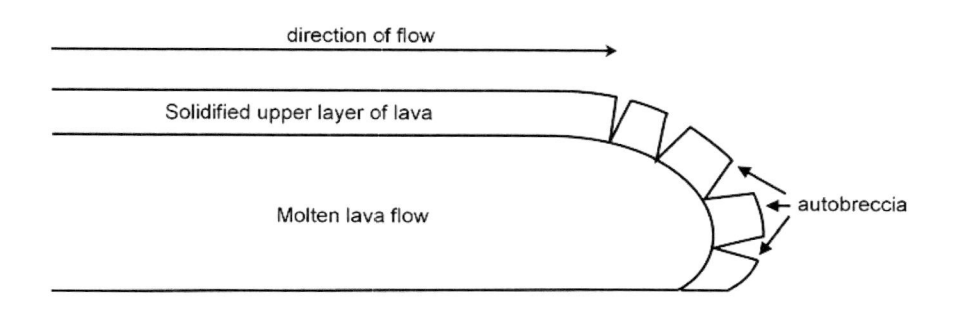

Fig. 5.7: Autobreccia forms at the front of lava flows as it rolls over.

another mile or so to another viewpoint where Hospital Canyon is below on the right and Limpia and Keesey Canyons on the left. Strata around the tower are in the first Barrel Springs unit, just above the tuff at its base.

Reset your odometer on returning to Highway 118. Mileages to the McDonald Observatory are measured from the park entrance.

## Stop 3: Sleeping Lion Formation (0.1-0.3 mile from State Park)

In a long road cut just north of the park entrance, Sleeping Lion lava at road level is an autobreccia, in which angular blocks of included lava are of the same material as their matrix (Fig. 5.8). Autobreccias form at the front of a lava flow where solidified material on the surface breaks up as the flow moves forward (Fig.5.7).

Channel deposits of tuff, sandstone and conglomerate occur above the lava. Erosion channels in the lava filled with debris such as pebbles and volcanic ash which in time solidified into the channel deposits. Above the channel deposits are layers of white and pink tuff from the initial eruption of the Barrel Springs Formation, the pink tuff sometimes being called the Fort Davis Tuff. Tuff units are from 3 to 30 feet thick.

Limpia Canyon becomes smoother beyond the tuff exposures with fewer rock outcrops, vegetated with junipers and other trees on its flanks on either side of the road. It broadens out on the right into an alluvial flat through which Limpia Creek meanders.

Rounding a quite sharp corner at Mile 0.9, a small waterfall comes down a crevice in a low jointed cliff across the creek. Continuous lavas are on display along this section. Several exposures of tuff at the base of the Barrel Springs Formation can be seen in road cuts on the left side of the road as it winds towards the Prude Ranch entrance (Fig. 5.9). Although imperceptible, the base of the formation has steadily dipped to the west at about 80 feet per mile from the State Park overlook.

## Stop 4: Prude & Sproul Ranch Entrances (1.6 miles from State Park)

The canyon opens out on the right as Limpia Creek veers sharply to meet Cook Creek a half-mile back from the road on the right. Casket Mountain (6,183 feet), at 3:30 o'clock from the entrance, is named for its shape. The mountain is capped by Casket Mountain lava, a porphyritic rhyolite very similar to the Mount Locke Formation and one of the youngest dated rocks in the Davis Mountains (35.2 Ma).

The Sproul Ranch, a working cattle ranch established in 1886 and owned by its founder's descendant, accommodates hunters and other visitors in 6 suite-style hotel rooms and a cabin. The family also owns the Harvard Hotel in Fort Davis. Its web site is www.sproulranch.com.

Prude Ranch runs a children's summer camp and is operated by the fifth generation of the Prude family. Among its accommodations are ranch bunkhouses, used by the children at camp, guest lodges and RV camp sites. Its web site is www.prude-ranch.com.

## Stop 5: Ramp Structure in Barrel Springs Formation (2.7 miles from State Park)

Beyond the Prude Ranch entrance the road runs along a hollow between a ridge on the right and the canyon wall on the left. An interesting example of a ramp structure can be seen in the left road cut. On the left side of the cut is an autobreccia that includes pyroclastic fragments of pumice and vitrophyre. Midway in the cut, a massive lava flow rode up over the autobreccia in a ramp. Flow bands in the lava show the path taken by the lava (Fig. 5.10).

## Stop 6: Limpia Crossing Entrance (3.4 miles from State Park)

About 3 miles from the park entrance, the canyon begins to open on both sides of the highway. The wonderful 270 degree panorama of the upper Limpia Creek drainage basin with its surrounding mountains comes into view at the Limpia Crossing main entrance.

On the left from the entrance, a grassy field slopes up from the road, a pastoral scene overlooked by Blue Mountain at 10:30 o'clock (Fig. 5.11). Paradise Mountain at 11:30 o'clock has a great light-colored fan of loose broken rock or talus below the summit (Fig. 5.12). At 12:30, the peak of Mount Livermore is on the skyline. Limpia Mountain is at about 1 o'clock with Pine Peak behind and to its right on the horizon.

Fig. 5.8: In this road cut at Mile 0.2 from the state park entrance, the top of the Sleeping Lion Formation is an autobreccia in which angular blocks of lava are included in a matrix of the same material. Above it are tuffaceous deposits that filled in channels in the lava. Above them are white and pink tuff beds from the initial eruption of the Barrel Springs Formation.

Fig. 5.9: Looking up Limpia Canyon from Mile 1.2. Limpia Creek meanders in the foreground with Prude Ranch buildings in the middle distance. Blue Mountain is on the skyline at left with Mount Livermore behind it, and Pine Peak on its right.

Fig. 5.10: Ramp structure in Barrel Springs strata at Mile 2.7 from the state park. Midway along the cut, lava flowing from the right overrode an autobreccia on the left. Flow banding in the lava show the path it took.

Fig. 5.11: Blue Mountain from Limpia Crossing entrance at Mile 3.4 showing the pastoral ambience of the Davis Mountains in autumn

Arabella Mountain at 2 o'clock is a ridge with smooth sides and rough rocky summit, sloping down towards Limpia Creek to the left. Just to its right, you can see the white Otto Struve and Harlan J. Smith telescope domes of the McDonald Observatory on Mount Locke (6,791 ft.) and the silver dome of the Hobby-Eberly telescope on Mount Fowlkes (6,659 ft.).

Olds Creek comes into the basin just below the domes and joins Limpia Creek on the right. At 3:30 o'clock, a high mesa (5,748 feet) is capped by a prominent Wild Cherry tuff ledge near its summit in which a fall has exposed some of the white tuff.

Beyond the entrance, the road descends to the valley of Limpia Creek below, with flat grassland to the left while the scattered houses of Limpia Crossing can be seen through the trees on the more wooded right. The highway crosses in quick succession the creek coming down Jones Canyon at Mile 4.6 and then Limpia Creek (Mile 4.9). Jones Canyon erodes back into the mountains to the right of Blue Mountain (7,286 feet) on the skyline at 9 o'clock.

The highway bears right as it turns north around Arabella Mountain past the entrance to the McIvor Ranch (Mile 5.7). Deer Mountain (6,126 feet) is just above the ranch buildings with Limpia Mountain (6,539 feet) on its right flank. An old solar energy demonstration site, set up and operated by AEP, the local power company on land donated by the McIvor Ranch, is no longer in use (Mile 6.1).

## Stop 7: Limpia Canyon Ranch (6.8 miles from State Park)

An oil exploration well, the Shell No. 1 McIvor, drilled about a mile up the ranch road in 1960, intersected 3,570 feet of volcanic strata before hitting Cretaceous and Permian dolomite and limestone at 1,786 feet above sea level. Nineteen miles due east, the Continental No. 1 Kokernot well intersected the base of the volcanic rocks at 3,248 feet, showing that the top of Cretaceous strata drops about 80 feet per mile to the west. This agrees with the earlier observation that the base of the Barrel Springs Formation tilted at 88 feet per mile to the west.

Mount Livermore (8,381 feet), the highest point in the Davis Mountains, is on the skyline at 10 o'clock, Pine Peak at 10:30 and Mount Locke at 1 o'clock with Mount Fowlkes just to its right.

Just beyond the entrance, the road bears right and begins ascending the lower slopes of Mount Locke through Barrel Springs Formation strata. Some are densely welded flows, others yellowish or cream-colored devitrified breccia with included angular lava blocks.

## Stop 8: Picnic Area (9.1 miles from State Park)

Halfway up the mountainside, a picnic area on the left overlooks the Limpia Creek drainage basin, 600 feet below (Fig. 5.13). Limpia Mountain is just before 9 o'clock in center front with Blue Mountain behind it at 8 o'clock. Paradise Mountain, in a caldera which is the probable source of the Wild Cherry and Mount Locke formations, is at 9:30 o'clock on the skyline. Across the highway, tuff crops out in beds dipping at between 15 and 20 degrees to the west, sandwiched between Sheep Pasture and Barrel Springs rhyolites (Fig. 5.15).

The base of the Barrel Springs Formation here is about 850 feet higher than just before the Prude Ranch, in both places tilting down to the west. How does one explain this structure? One possibility is that we are near the edge of the Paradise Mountain caldera, first mapped by Henderson in 1989. He proposed that the eastern margin of the caldera was in the vicinity of the Davis Mountains State Park, and that its northern boundary ran along Limpia Canyon so perhaps the highway has been in the caldera from the State Park and has now climbed up the margin. Sporadic faulting along the escarpment from Paradise Mountain supports this possibility.

Another factor to be considered is that we are on the eastern edge of the Brooks Mountain-Mount Livermore dome (Fig. 2.14) which may have uplifted Barrel Springs strata here. A third factor is that we are near the limit of Sheep Pasture Formation lava so conceivably the overlying Barrel Springs flows may have spilled over the edge of the Sheep Pasture Formation, at least 165 feet thick on Mount Locke and 560 feet thick on Sheep Pasture Mountain (see Fig. 5.14). See also Chapter 9 for further discussion of the structure of the mountains.

Coming back to the strata on display across from the picnic area, the 170-feet thick section of tuff is described by Mattison as consisting of 21 separate flows deposited as ash flows, poor to moderately welded. They are mostly glass or devitrified glass and flattened yellowish-green pumice fragments (called fiamme by geologists) with some quartz and feldspar crystals. No stream deposits and conglomerates are present so they have not been reworked. The overlying lava has baked and hardened the upper tuffs, the heat oxidizing the tuffs and turning them red.

The Barrel Springs unit consists of, from the bottom up, a basal breccia, 2 feet thick, 110 feet of reddish, porphyritic, platy rhyolite, with columnar jointing, and 165 feet of an upper flow breccia. The basal breccia includes tuff fragments at its base, and is flow-banded. Midway along the outcrop is a near-vertical laminated weathered dike.

Fig. 5.12: Looking across to Paradise Mountain on the horizon at right from Limpia Crossing entrance at Mile 3.4. An irregular escarpment stretches from Paradise Mountain to Brown Mountain and White Mountain, behind Brown Mountain on the left. The escarpment is capped by Goat Canyon rhyolitic ash-flow tuff, overlying tuffs of Wild Cherry. The great fan-shaped slope of Goat Canyon boulders on Paradise Mountain is called a talus slope by geologists.

Fig. 5.13: Upper Limpia Creek basin from the picnic area at Mile 9.1. Limpia Mountain is on the far left. The tip of Paradise Mountain is on the horizon at far right. Limpia Creek runs to the right of Limpia Mountain and in front of Paradise Mountain.

# 5: FORT DAVIS – KENT

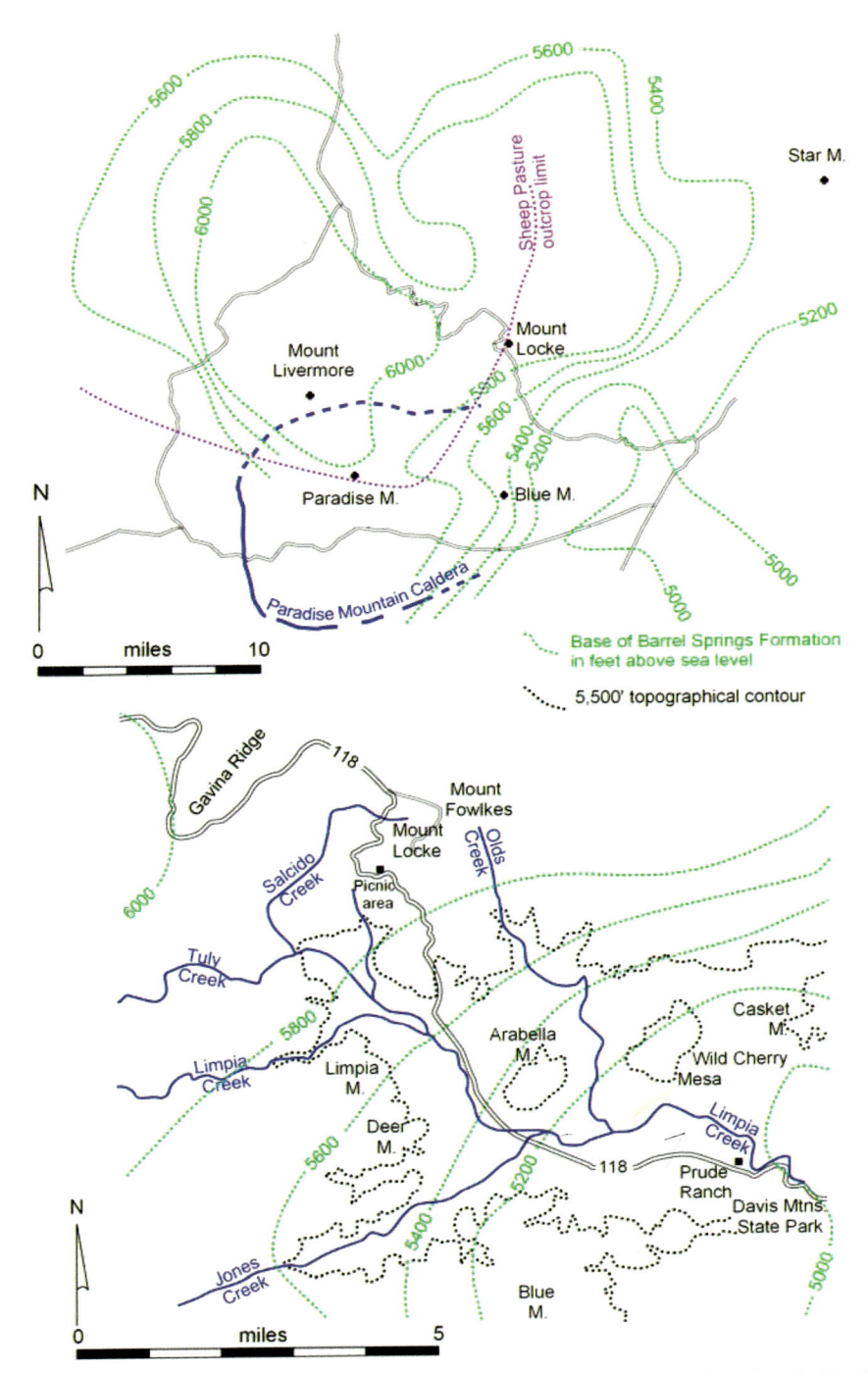

Fig 5.15: The structure of the Barrel Springs Formation above with Limpia Basin detail below.

Round the curve, yellow-green tuffaceous mudstone and sandstone overlain by bedded tuffaceous materials, up to 20 feet thick, lie between the Barrel Springs strata and Mount Locke trachyte. The latter is 625 feet thick here and consists of multiple flows, each up to 40 feet thick. The formation came from a source in the Paradise Mountain caldera where it is more than 800 feet thick. It crops out on the right almost continuously to the pullout at Mile 10.3, pitted and coarsely porphyritic. Alkali feldspar phenocrysts up to an inch long make up as much as 40 per cent of the rock.

### Stop 9: Deadman's Canyon Pullout (10.3 miles from State Park)

The highway continues up to a summit at Indian Hill and then follows Deadman's Canyon to a pullout and sharp corner around a dry stream at the head of the canyon. The canyon got its name when Horace Powe's dead body, riddled with eleven bullets, was found propped up against a boulder in the canyon in January 1884, three weeks after he had disappeared before a dance in Fort Davis. His murderer was never caught; a neighboring rancher named Brown disappeared around the same time. Barry Scobee tells the story in his booklet *"The Steer-Branded Murder"*, republished by the Fort Davis Historical Society in 2004.

### Stop 10: McDonald Observatory (10.9 miles from State Park)

From the pullout, the road climbs a slight incline around Mount Locke lava ridge to the entrance to the McDonald Observatory Visitor Center. The main entrance is 200 yards ahead. Administered by the University of Texas at Austin, the observatory was built with the proceeds of a bequest in 1929 by lawyer and banker W. J. McDonald of Paris, Texas. It sits on 650 acres of land on Mount Locke donated by the McIvor ranch. The big domes you see as you approach from the entrance are, from left to right, the silver 433-inch Hobby-Eberly telescope on Mount Fowlkes (6,659 feet) and the white 107-inch Harlan J. Smith and 82-inch Otto Struve telescopes on Mount Locke (6,791 feet).

The Hobby-Eberly telescope, which began operations in 1999, is one of the largest optical telescopes in the world. The larger the mirror, the fainter emissions it can detect and hence the farther into space it can penetrate. It is a joint project of The University of Texas at Austin, Pennsylvania State University, Stanford University, Ludwig-Maximilians-Universität München, and Georg-August-Universität Göttingen. The building has a visitors' gallery with television commentary. Visitors can see the reflecting telescope through glass walls.

The observatory also operates a 30-inch telescope and a laser system that measures the distance between Earth and the Moon and tracks the drift of the Earth's continents. It produces daily astronomy radio programs in English, Spanish and German and StarDate magazine (see http://stardate.org). The visitor information center is open to visitors daily from 9:00 am to 5:00 pm, offering solar viewing sessions and guided tours. Every Tuesday, Friday, and Saturday beginning at evening twilight, visitors have the opportunity to view the planets, moon, galaxies, and other celestial objects through amateur telescopes. Times and fees for guided tours and Star Parties can be found at http://mcdonaldobservatory.org.

Note the lichen-covered boulders of Mount Locke porphyritic trachyte followed by pinkish foliated Wild Cherry tuff on the road up to Mount Locke. Lichen is rare in Texas except in the cooler high altitudes of the Big Bend. The summits of both Mounts Locke and Fowlkes provide fine views to the east and south (Fig. 5.17).

## Stop 11: Eppenauer Ranch (0.4 miles from Observatory)

Reset your odometer as you leave the observatory. Beyond the observatory entrances, the highway crosses a level meadow studded with junipers and live oaks. The meadow is on a drainage divide being eroded by Little Aguja Canyon, three miles away on the right and draining to the north, and Salcida Canyon, a half-mile to the left in the Limpia Creek basin and draining to the east (Fig. 9.6).

The pyramid-shaped Guide Peak (6,840 feet) is at 4:30 o'clock from the ranch entrance with Spring Mountain (6,965 ft.) to its left, both with Wild Cherry tuffs above Mount Locke lava. Gavina Ridge (6,540 feet) is at 12 o'clock with the flattish topped Brown Mountain (7,274 ft.) to its left. Blue Mountain is conspicuous on the horizon at about 8:30. The Marfa aerostat (see page 177) can sometimes be seen at 9 o'clock. The Eppenauer Training Stables just beyond the ranch entrance accommodated and trained quarter horses in the summertime for many years but are no longer in use.

## Stop 12: Gavina Ridge Water Tank (2.6 miles from Observatory)

Three hundred yards ahead, the road bears sharply left and winds along the base of Gavina Ridge at about 6,270 feet elevation, the highest point on Highway 118. There are few outcrops except in road cuts although the ridge has some light-colored Mount Locke lavas cropping out on its summit. At the water tank, a magnificent panorama can be seen on the skyline at 9 o'clock through a gap between Mount Locke and Blue Mountain. On the far horizon, the Glass Mountains are just to the right of Mount Locke, followed by Bird Mountain. Mitre Peak is at about 9 o'clock with Mount Ord on its right shoulder and Twin Peaks just before the slopes leading up to Blue Mountain.

Fig. 5.15: This sequence of dipping tuffs at the picnic area, Mile 9.1, lies between brecciated gray Sheep Pasture rhyolite on the right and Barrel Springs foliated rhyolite, brecciated at base, on the left. The latter baked and reddened the upper tuff beds.

Fig. 5.16: Foliated and heavily altered dike in the Barrel Springs Formation just uphill from the above photograph. It is also exposed on the other side of the ridge about a mile ahead.

Fig. 5.17: Looking towards Alpine from Mount Locke. On the horizon, Mount Ord is on the far left followed by Mitre Peak, Twin Peaks, and the Haystacks bracketing Cathedral Mountain. The Puertacitas Mountains are about the center of the horizon with Goat Mountain to their rear on the left. Arabella Mountain is in mid-center with Blue Mountain on the far right.

Fig. 5.18: The Nature Conservancy's Davis Mountains Preserve.

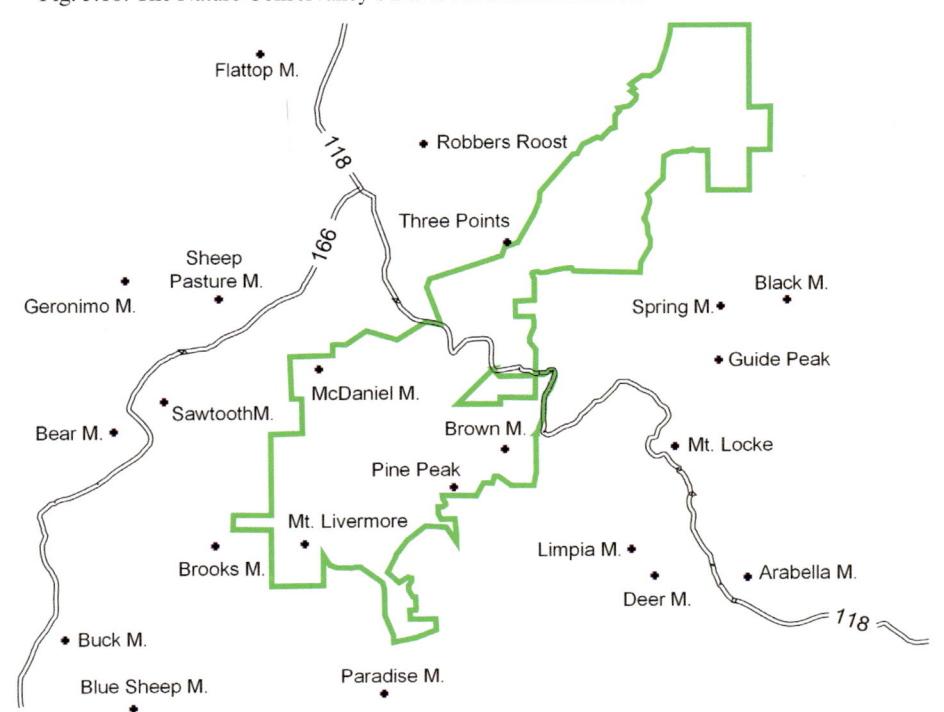

Brown Mountain (7,273 feet), a ridge of Mount Locke lava, is on the left as you turn around the end of Gavina Ridge. The road then winds downhill along Brown Canyon between the ridges, crosses the canyon at Mile 4.5 and turns sharply left to follow Elbow Canyon at Mile 5.4. The hills are rounded and quite well vegetated with grass and scattered shrubs. Live oak, juniper, and pine trees line the highway.

At the flood gauge on Elbow Canyon (Mile 6.0), the highway crosses a fault hidden by alluvium in the canyon floor that runs parallel to the road for the next mile or so, displacing strata up to the right, so that Barrel Springs flow autobreccia crops out at road side uphill from the flood gauge.

Higher up towards the saddle called Fisher Hill (6,119 feet), the highway intersects scattered debris and small outcrops of red and yellow tuffaceous material at the base of the Mount Locke Formation, similar to that seen at the picnic area half-way up Mount Locke (page 77). Along this stretch, Pine Peak (7,736 ft.) is on the skyline a little after 9 o'clock.

Just beyond the crest, porphyritic Mount Locke lava is again juxtaposed against much less porphyritic Barrel Springs lava by more faulting on the right. White fault breccia can be seen along the fault contact. Below that, rubble and small, finely bedded, yellowish gray tuff outcrops appear at roadside for 200 yards followed by Mount Locke porphyry. At the bottom of the hill, the road crosses Madera Canyon creek (Mile 7.8). The creek valley on the right is a beautiful flat-bottomed grassy stretch studded with quite sizeable oak and pine trees.

## Stop 13: Madera Canyon Roadside Park (8.2 miles from Observatory)

Madera Canyon Roadside Park, officially the Lawrence E. Wood Roadside Park, is on the right about a quarter-mile beyond the creek crossing, wooded with ponderosa pines. Note the heavy lichen growth on the jointed cliffs on the other side of the stream.

The land between Madera Canyon and Mount Livermore, including the mountain itself, belongs to The Nature Conservancy, a non-profit organization. In July, 2004, the Conservancy announced it had purchased 10,000 acres from the Eppenauer family and trust. This block of land, the Eppenauer's Fisher ranch, links the 18,000-acre Davis Mountains Preserve and the 4,000-acre Madera Canyon Preserve, bringing the total of Conservancy-owned preserve land in the Davis Mountains to 32,000 contiguous acres (Fig. 5.18). A further 70,000 acres are owned by conservation partners of the organization who have conservation easements to preserve the land.

## Stop 14: Davis Mountains Preserve entrance (8.9 miles from Observatory)

The road runs along the canyon floor for a quarter-mile beyond the picnic area and then turns sharply right. One of the best views of Mount Livermore is directly down the access road at the Davis Mountains Preserve entrance on the left (Fig. 5.19).

The Conservancy opened the McIvor Conservation Center, a visitor and research center, a short distance down this road in November, 2004. The Conservancy plans to construct an interpretative nature trail loop to take visitors across Madera Creek and up through pinion-oak-juniper woodlands, where hikers will be able to enjoy views of Mount Livermore and the ponderosa pines of Madera Canyon.

## Stop 15: Beef Pasture Gap (9.5 miles from Observatory)

Beyond the Davis Mountains Preserve entrance the highway climbs up gradually to Beef Pasture Gap through rounded hills with scattered shrubs. Trees at roadside include oak, juniper and pinon pine. Nobbly Barrel Springs lava is on the left, tuffaceous in places. Several hundred feet of jointed Wild Cherry, Mount Locke and Barrel Springs strata crop out in a high bluff (6,644 ft.) on the left above the saddle. Mount Locke lava caps a similar bluff (6,440 feet) on the right above a small canyon.

The summit of the gap, not an especially steep one, is at 6,185 feet, about the same elevation as the Observatory entrance. From this point on, the highway descends some 500 feet down the north front of the Davis Mountains to the junction with Highway 166.

Erosional knobs on both sides of the road are capped by Barrel Springs strata overlying Sheep Pasture rhyolite. The highest such remnant is McDaniel Mountain (7,235 feet) at about 9:30 o'clock on the horizon from Mile 10.2 (Fig. 5.20).

## Stop 16: Caldwell Ranch (12.3 miles from Observatory)

The canyon opens out a mile below the summit into attractive rolling grassland ringed with high juniper-studded mountains. The Caldwell Ranch is part of the current 70,000 acres that the Nature Conservancy has under conservation easements, having bought it from its previous owners and resold it to a conservation partner.

The very rocky Robbers Roost Mountain (6,652 ft.) at 1:30 o'clock from the entrance (Fig. 5.21) and Three Points Mountain (6,562 feet) just after 3 o'clock are other examples of Barrel Springs outliers overlying Sheep Pasture steep-sided cliffs.

Fig. 5.19: Mount Livermore from Nature Conservancy entrance at Mile 8.9.

Fig. 5.20: McDaniel Mountain from the gate at Mile 10.2. The mountain is capped by Barrel Springs overlying Sheep Pasture strata.

Fig. 5.21: Robbers Roost from Caldwell Ranch entrance at Mile 12.3. The jointed cliffs are of Sheep Pasture rhyolite overlain by craggy Barrel Springs strata.

Fig. 5.22: The plain beyond the mountains from the summit of Nunn Hill, Mile 13.1. Flattop Mountain is in the center middle distance. Adobe Canyon Formation lava flows make up the long ridge on the skyline. The skyline descends into Adobe Canyon on the extreme right of the photograph.

Two hundred yards beyond the ranch entrance, Sawtooth Mountain comes into view on the horizon at a little before 9 o'clock, spectacularly rugged.

As you descend to the intersection, the quite rounded mountain slopes with few rock outcrops are the result of the higher rainfall here. Prevailing winds in this part of Texas are from the north and west and here are intercepted by the mountains and driven up. Any water vapor is condensed by the lower temperatures and descends as rain. The higher rainfall creates more erosion, more soils and grasses, and more rounded slopes.

### Stop 17: Summit of Nunn Hill (13.1 miles from Observatory)

The steep-sided little butte on the right as the highway approaches the saddle at Nunn Hill (5,925 ft) is an outlier of Barrel Springs over Sheep Pasture strata. Road cuts at the summit expose purplish Sheep Pasture rhyolite lava blocks in a yellowish gray clayey matrix. Nelson and Nelson describe it as a devitrified vitrophyre overlain by a densely welded ash flow, oxidized to a reddish color. The sharp pyramid at 12:30 o'clock (6,412 feet) is another Barrel Springs outlier.

Fine views of the terrain ahead can be seen from the summit. Most of the nearby hills, including Flattop Mountain, front center, are outliers of Sheep Pasture rhyolite overlying Frazier Canyon and Adobe Canyon strata. On the horizon, Adobe Canyon cuts through the long ridge of Adobe Canyon lavas (on the extreme right of Fig. 5.22).

### Highways 118/166 Junction (14.0 miles from Observatory)

Reset your odometer at the second entrance to Highway 166.

For the next six miles, the highway runs through the ragged northern edge of the Davis Mountains at about 5,400 feet above sea level. The highway is on Frazier Canyon Formation which here underlies the Sheep Pasture Formation. It seldom crops out and is seen only in scattered road cuts as weathered reddish brown lava and tuff. At Mile 2.0 and again at Mile 2.2, Frazier Canyon tuff with inclusions and red and gray lavas are exposed in road cuts on the right.

At Mile 3, Flattop Mountain (6,023 feet) is on the left with a butte (6,003 feet) across the highway, its Sheep Pasture lava cap dipping ahead. Both are outliers of Sheep Pasture overlying Frazier Canyon and Adobe Canyon Formations. At this point, the highway starts slowly downhill through grassland with scattered cholla and yucca shrubs and the occasional tree in the lower ground.

## Stop 18: Adobe Canyon Road Cut (4.1 miles from 118/166 Junction)

The first close-up of the Adobe Canyon Formation is in a road cut on the right, lava altered to clay overlain by unaltered lava. The formation covers nearly 500 square miles in the north and northeast Davis Mountains, second only in area among lavas to the Star Mountain unit (Fig. 2.12). It consists of up to 3 flows of slightly porphyritic lavender-gray rhyolite weathering reddish-brown to dark brown, each up to 600 feet thick. Each flow has basal and upper breccias. Generally, each will form a separate cliff in an outcrop, often with columnar jointing at top; some have been traced for 6 to 8 miles. Occasionally, zones of tuff and volcaniclastic material occur between flows, seen as lighter-colored hollows on hillsides where they have been eroded.

The thickest flows occur along the south and southwest margins of the Buckhorn caldera and one or more of the sources may be in that vicinity. Kathleen Childerrs, who studied the unit for her Baylor University B.S. thesis, suspected that one source of the lava might have been near Bear Mountain, about 8 miles east of here, where it is thickest (1,040 feet). If two centers, the other would be between Boracho Canyon and Herds Pass about 6 miles west of Round Mountain, 10 miles ahead. Three flows are found there, 800 feet thick in total.

A half-mile ahead, Friend Mountain (6,218 ft.) is on the right across attractive grassland with scattered shrubs, especially cholla. Its prominent Sheep Pasture lava bed is above a Frazier Canyon escarpment.

## Stop 19: Hi-Lonesome Pens (5.3 miles from 118/166 Junction)

Just beyond a Caldwell Ranch entrance on the right, Adobe Canyon Formation lava is exposed in road cuts on both sides of the highway, somewhat bedded, somewhat foliated, somewhat folded, somewhat faulted. Similar exposures occur 200 yards ahead, just beyond the Hi-Lonesome Pens road on the left, where foliated and brecciated lava can be seen in road cuts on both sides of the highway.

## Stop 20: Adobe Canyon (5.8 miles from 118/166 Junction)

The highway enters Adobe Canyon here (Fig. 5.23). The canyon is narrow and flat bottomed, about 200 yards across in its upper reaches with smooth, rounded sides covered mainly by thin vegetation over pebbly soil. Some live oak and mesquite are seen at watercourses but otherwise the canyon is not heavily wooded. Outcrops of jointed lava appear spasmodically and as foliated and folded exposures in road cuts.

Fig. 5.23: Entrance to Adobe Canyon from Mile 4.0. The hills in the canyon are more rounded than usual in the Davis Mountains. The northern face of the Adobe Canyon Formation in Fig. 5.24 is rugged by comparison.

Fig. 5.24: Encantada Peak, on the left, and Round Mountain from Mile 15.9. Two flows of Adobe Canyon rhyolite can be seen to the right of the photograph with a light tuff outcrop in the break between flows on the second promontory. A third flow is on the summit of Encantada Peak near the left margin.

Fig. 5.25: Horse Camp Peak, seen here from Mile 18.4, west of Encantada Peak, has three flows of Adobe Canyon rhyolite overlying Gomez and Huelster tuffs. The cliff is formed by the lowest Adobe Canyon flow.

Fig. 5.26: Thinly-bedded sandy limestone of the Cretaceous Boquillas Formation is exposed in a road cut at Mile 19.5. Thin brownish sandstone layers are interspersed with pinkish cream biomicritic limestone, created from carbonate mud and skeletal debris.

The highway cuts 500 feet down through the Adobe Canyon Formation as it goes north and the lava in road cuts becomes blockier and less foliated. The hilltops are up to 6,250 feet on either side, 1,400 feet above the road, all of Adobe Canyon Formation.

At Mile 14.2, the road turns left with outcrops for the next half mile or so on the left of blocks of broken-up, jointed gray lava caught up in a clay matrix, some in place, some not. The road turns right with light-colored weathered tuffaceous material in the right road bank.

## Stop 21: Exit from Adobe Canyon (14.8 miles from 118/166 Junction)

At the canyon exit, the road turns right as the canyon opens. On the left an outcrop of gray-green and purplish brown lava blocks in a gray fine-grained clay matrix is at the base of the Adobe Canyon Formation above Gomez Tuff. The perimeter of the Buckhorn caldera crosses the road here and disappears under Adobe Canyon lavas (Fig. 5.1).

## Stop 22: Rancho del Cielo Entrance (15.9 miles from 118/166 Junction)

The Rancho del Cielo entrance on the left is at 4,508 ft. above sea level, 600 feet below the entrance to Adobe Canyon. The ranch house is set at the foot of cliffs between Encantada Peak (5,787 ft.) on the left and Round Mountain (5,846 feet) on the right. Both have caps of Adobe Canyon lava over Gomez Tuff with Huelster Formation tuffs at base (Fig. 5.24). Rock House Canyon runs into the mountains to the left of the ranch house.

The highway at the ranch entrance, nearly 300 feet below the ranch house, is on alluvium overlying Cretaceous strata, below the base of the volcanic rocks. Gomez Peak at 3 o'clock on the horizon is the northernmost peak in the Davis Mountains.

At Mile 17.6, you can see the Adobe Canyon lava escarpment from Encantada Peak to Horse Camp Peak (5,892 ft), the northernmost point of the Davis Mountains west of Highway 118 (Fig. 5.25). The peak consists of three Adobe Canyon lava flows 800 feet thick overlying Gomez and Huelster tuffs. The lower Adobe Canyon flow forms the striking cliff two-thirds of the way up the escarpment.

## Stop 23: Boquillas Formation (19.5 miles from 118/166 Junction)

In a road cut at the crest of a ridge, Cretaceous rocks are on view for the first time on this route, orangey brown, finely bedded sandy limestone of the Boquillas Formation (Fig. 5.26). This formation developed on an open shelf of the sea in which fine-grained sediments were deposited from time to time, creating the thin sandy layers. The sediments came from sources that were either a long way away or were from relatively flat-lying terrain. The limestone is a biomicrite, derived from a combination of skeletal debris and carbonate mud.

Going downhill from the summit, light brownish gray nodular Buda Limestone appears in road cuts on both sides of the road, underlying the Boquillas here. It was deposited on an open shelf in somewhat deeper water than the Boquillas. It is also a biomicrite, recrystallized in places.

## Stop 24: Highway 118/Interstate 10 Junction (22.5 miles from 118/166 Junction)

The road descends to Interstate 10 through limestone buttes and road cuts in the Boracho Formation, which underlies the Buda Limestone around the northern mountains. The Boracho is a clayey limestone with some thin shale beds that become more common towards the base. It contains abundant large marine fossils such as pelecypods, ammonites and echinoids (see Fig. 10.2 for the Cretaceous succession here).

Kent, like so many Trans-Pecos settlements, began as a shipping and watering point on the railroad. The Kent Mercantile Building is partly constructed from blocks of green, flow-banded Gomez Tuff.

Your options here are to turn right on Interstate 10 to Toyahvale, 30 miles away, or left to Van Horn, 27 miles to the west. The Toyahvale option takes you along the northern escarpment of the Davis Mountains, as described in Chapter 10, and then up to Fort Davis through Wild Rose Pass and Limpia Canyon, described in Chapter 4.

If you have already traveled the Toyahvale-Fort Davis leg, the Kent-Van Horn-Valentine-Fort Davis loop is an attractive option, particularly for the fine mountain scenery around Van Horn, 27 miles to the west, and for the excellent perspectives on the south and west Davis Mountains as you pass Valentine. This route is described in Chapter 10.

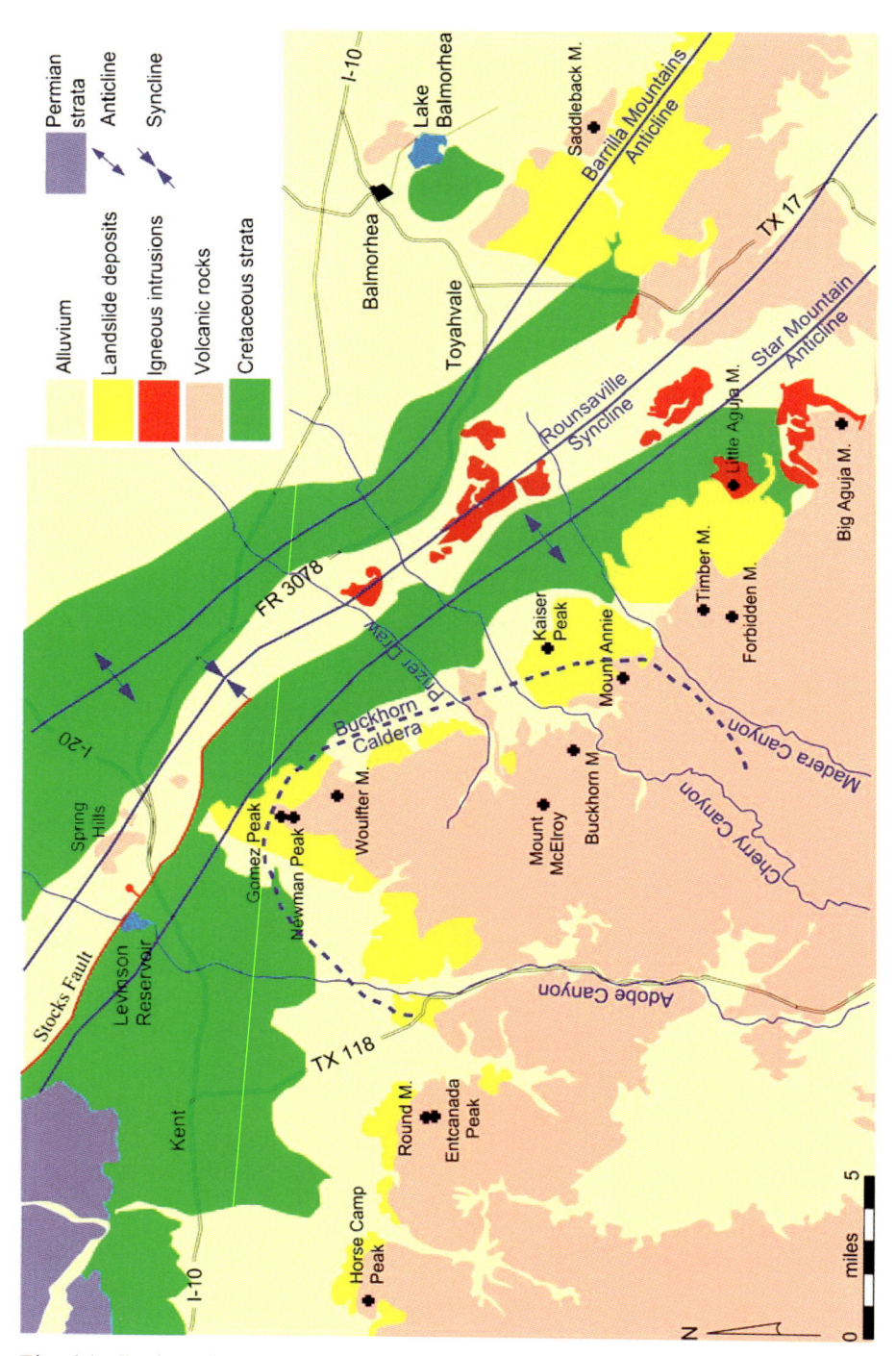

Fig. 6.1: Geology from Kent to Toyahvale

Fig. 6.2: The sharp Gomez Peak (6,320 feet) rises 2,280 feet above Giffen exit at Mile 11.0. The boulder-covered slope below the peak is called a talus slope by geologists.

Fig. 6.3: The northern section of the magnificent panorama from Gomez Peak to Star Mountain from the interstate at Mile 17.5. Woulfter Mountain is on the left of the photograph. Gomez Peak is on the right with Newman Peak just behind it. Level Cretaceous limestone beds crop out on the rounded hills in the middle distance. Huelster tuff between the Cretaceous strata and the Gomez Tuff of Gomez Peak crops out in the white patches to the right above the limestone.

## 6: Kent – Toyahvale

### 30 miles

At the intersection of TX 118 and Interstate 10, turn east on to the interstate at Mile 176.3, and reset your odometer. Mileages are measured from the intersection.

The interstate skirts the northern edge of the Davis and then the Barilla Mountains for the next 30 miles. The view for the first 15 miles is dominated by the high Gomez Tuff ridge on the right, which terminates with Gomez Peak (6,320 feet) at 1 o'clock and Newman Peak (6,302 feet) to its right.

At road level, on leaving Kent the interstate is in moderately bedded, grey and yellowish to brownish gray Boracho Limestone which is exposed in deep road cuts at Mile 1.5. Buda Limestone forms cliffs on top of the south-dipping cuesta on the right. Another road cut is at Exit 181 and once through it, you can see across open country to the volcanic hills in front right.

The axis of the Star Mountain anticline (Fig. 6.1) crosses the interstate about Mile 4.5. At Mile 5.7, Levinson Reservoir, created by damming Adobe Draw, is at 9 o'clock. Water flows into the quite marshy country between the lake and the interstate from limestone strata on the northern flank of the anticline.

### Stop 1: Spring Hills Road Picnic Area (8.2 miles from Kent)

The Stocks Fault, which runs along the northeast margin of the Apache Mountains, comes across the interstate at about the Spring Hills Road underpass. The incline coming down to the exit and a low valley on the right are the only indications of its presence. It displaces strata 1,620 feet down to the northeast near the Levinson Reservoir, where Buda Limestone is juxtaposed against Cox Sandstone. The fault continues on the right for several miles before dying out. Folding and faulting in this part of Texas mainly follows a northwest-southeast trend.

| Period | Age m.y. | Formation | Description |
|---|---|---|---|
| Upper Cretaceous | 98-45 | Undivided | Marl, shale and clayey limestone; 300 ft. thick. |
| | 144-98 | Boquillas | Upper part, interbedded marl and shale; lower part, limestone, silty to sandy, flaggy, dark grayish orange near base; marine megafossils; 200 ft. thick. |
| Lower Cretaceous | | Buda Limestone | Limestone, thin to thick bedded, middle 60 ft. clayey; sandstone locally at base; 140 ft. thick. |
| | | Boracho | Two members; *San Martine Limestone:* clayey limestone, thin to thick bedded with calcareous shale interbeds; marine megafossils; 230 ft. thick. |
| | | | *Levinson Limestone:* Upper two-thirds limestone, light to light olive gray with interbeds of yellowish marl and shale; lower third shale, light to dark gray with occasional limestone beds; abundant marine megafossils; 150 ft. thick. |

Fig. 6.4: Cretaceous strata along Interstate 10, north of the Davis Mountains. A megafossil is one large enough to be studied without the use of a microscope.

The dark, rounded Spring Hills at 11 o'clock from the picnic area are composed predominantly of Gomez Tuff, where it is at its thickest (425 feet) outside the Buckhorn Caldera.

## Junction of Interstates 10 and 20 (9.6 miles from Kent)

At the junction of the two interstates, Interstate 20 goes northeast between limestone hills on the left. Keep right on Interstate 10.

After passing an outcrop of quite blocky brownish-cream Buda Limestone, Gomez Peak is narrow, almost bullet-shaped at 3 o'clock from the Giffen Road exit (Fig. 6.2, Mile 10.8). Huge numbers of boulders litter the slopes below. Woulfter Peak (6,395 feet) is on its right.

The highway cuts through limestone ridges of brick-red finely bedded Boquillas Formation at Mile 12.8 and the thicker bedded Buda Limestone at Miles 13.5 and 13.8.

## Stop 2: Farm Road 3078 Exit (15.7 miles from Kent)

Take this exit to Toyahvale and Balmorhea. Before the interstate was built, FR 3078 was a busy road on one of the main east-west arteries across the southern United States. Now it is practically unused except for local traffic. However, it remains one of the best ways to view the unrivaled panorama of the northern Davis and Barilla Mountains which stretches from Gomez Peak at 2:30 to Star Mountain at 11 o'clock.

Fig. 6.5: The profile from the Barrilla Mountain ridge anticline on the left skyline to the Rounsaville syncline forming the low ground near the left to the Star Mountain anticline on the right from Mile 20.8, just before Cherry Creek. The Barrilla Mountains anticline is breached i.e. its crest has been eroded, where the profile dips midway across the ridge. The Star Mountain anticline is approximately at the summit of the Star Mountain. All three folds run towards the camera parallel to the highway.

Fig. 6.6: A panoramic view across the plain to the flat-topped Forbidden Mountain on the left, the sharp pyramid of Little Aguja Mountain in center and Timber Mountain on the right. The brown outcrops in the middle foreground are of the Barrel Springs lower ash-flow tuff, brought down to this level by the Rounsaville syncline, whose axis runs across the photograph in front of Little Aguja Mountain.

For its first 10 miles, FR 3078 runs between the Rounsaville Syncline and the Barrilla Mountains anticline to the left. The latter has created a low ridge that the highway crosses at Mile 25.2. These folds, the Rounsaville syncline and the anticlines on either side of it, run southeast across the mountains for about 60 miles or so, and are a major factor in the topography south of Toyahvale (see Chapter 4, Fig. 6.5).

Over much of its area, the Gomez Tuff crops out as a thin bed, 15-30 feet thick, but it thickens to an average of 450 feet in the Buckhorn caldera and almost 1,500 feet on Gomez Peak. The boundary of the caldera comes down Madera Canyon just to the right of Timber Mountain, the flat-topped mesa at 1 o'clock, across the low ground in front and disappears from view north of Gomez Peak at 4 o'clock (Fig. 6.1).

The Barrilla Mountains anticline ridge is on the left as you leave the interstate. Cretaceous Buda Limestone has been brought up to the surface by the anticline although outcrops are rare except for a small quarry in limestone at Mile 18 on the left.

The highway goes gently downhill in the fourteen miles from the freeway exit to the Toyahvale junction with Highway 17, crossing several major water courses along the way, such as Prizer Draw at Mile 18.6, running up the right of Mount McElroy (6,537 feet) at 3 o'clock, and Cherry Creek whose broad gravel-bottomed dry bed crosses the road at Mile 21.4. Vegetation here is mainly catclaw and mesquite.

Just beyond Cherry Creek, the road veers to the left to avoid the Red Hills, a series of intrusive sills on the right. It crosses the county line at Mile 25.2 and climbs a rise over the Barrilla Mountains anticline ridge. On the other side it descends to the Toyah basin, an alluvial basin that underlies most of Reeves County from Toyahvale to Pecos, in which alluvium is up to 1,500 feet thick.

The anticline ridge is roughly on the boundary between the Diablo Platform and the Permian Delaware Basin, and the sag in the Earth's surface that produced the Toyah basin is perhaps due to continuing development of the Delaware Basin.

At Mile 26.8, a low adobe structure and some wooden ruins on the right are all that is left of a motor court that, according to Tom Johnson, ranger at the Balmorhea State Park, hosted the bank robber John Dillinger and his gang in the 1930s. They apparently practiced shooting their firearms in their spare time.

At Mile 30, the junction with Highway 17 marks the end of this segment. Balmorhea State Park, just beyond the junction, is described in Chapter 3. To continue to Fort Davis, turn to Chapter 4.

# 7: Davis Mountain Scenic Loop

## 44 miles

Highway 166 branches off Highway 118 29 miles north of Fort Davis and loops around the west and south of the Davis Mountains to connect with Highway 17 just south of Fort Davis. It provides excellent views of the high mountains around Mount Livermore and Brooks Mountain from the west and south, and of the interior of the Paradise Mountain caldera.

The geology along the loop is given in Fig. 7.2. The various formations seen along the loop are described in Fig. 7.1. Set your odometer at the junction. Mileages in this segment are measured from there.

The loop follows Adobe Draw for the first 6 miles. The draw is one of those mysterious valleys that are difficult to explain. The innocuous and mostly dry stream that now runs down the valley has, at some time in the past 30 million years, carved a path nearly 1,500 feet deep and 6 miles long through hard Sheep Pasture lavas. Further north, the same stream created Adobe Canyon, 1,500 feet deep and 9 miles long, through the even harder lavas of the Adobe Canyon Formation.

The draw begins on the flanks of Whitetail Mountain ahead on the left and is quite straight until turning north just west of the junction. Therefore it seems likely that it follows a fault or fracture zone now covered by alluvium that provided a conduit for erosion. Adobe Canyon is not straight and is most probably the work of a stream that had set its course before eroding down to the Adobe Canyon Formation.

At Mile 3, the summit of the ridge at 3 o'clock is Sheep Pasture Mountain (6,950 feet). Sawtooth Mountain (7,686 feet) dominates the horizon in front along this stretch. Note how the slopes on the right have only scattered shrubs while the ones on the left are heavily wooded. Biologist John Karges of The Nature Conservancy explains that the slopes on the right face southeast and are more exposed to sunlight than those on the left. They thus retain less moisture and are less conducive to tree growth for woodlands than for grasses and shrubs.

# 7: DAVIS MOUNTAINS SCENIC LOOP

| Unit | Age m.y. | Description |
|---|---|---|
| Mount Livermore volcanic dome | undated | Trachyte volcanic dome; up to 1,400 feet thick on summit of Mount Livermore. |
| Brooks Mountain Formation | undated | Porphyritic trachyte volcanic dome; up to 985 feet thick on Brooks Mountain. |
| Goat Canyon Formation | 35.3 | Rhyolitic ash-flow tuff, 200 ft. thick on Brown M.; source unknown. |
| Wild Cherry tuffs |  | Fine-grained rhyolite tuff, up to 355 ft. thick, produced from Paradise M. Caldera in eastern Davis Mountains. |
| Casket Mountain lavas |  | Up to 5 flows of porphyritic rhyolite capping ridges in southern, central and eastern Davis Mountains; 1,600 ft. thick on Blue M.; probably erupted from widely spaced fissures. |
| Mount Locke Formation | undated | Quartz trachyte & rhyolitic porphyry in central Davis Mountains, gray when fresh to brownish gray to reddish brown when weathered; maximum thickness 580 ft.; source unknown. |
| Tuffaceous sediment, mafic lava |  | Tuffaceous sediment & mafic lava are at same horizon as Mount Locke Formation in Blue M. and Musquiz Canyon in southeastern Davis Mountains; source unknown. |
| Barrel Springs Formation | 35.6 | Upper unit: rhyolitic lava; source as for lower unit. |
|  |  | Lower unit: strongly rheomorphic ash-flow tuff or lava in central and eastern Davis Mountains; source presumably buried in central Mountains; up to 500 ft. thick. |
| Merrill Formation | undated | Porphyritic trachyte lava, 330 ft. thick south of Barrel Springs Ranch; probably produced around Paradise Mountain caldera. |
| Lower Basalt | undated | Basalt, dark gray to black, consists of several flows up to 80 ft. thick; found around Barrel Springs Ranch; source unknown |
| Sheep Pasture Formation | undated | Rhyolite, grayish purple to brown, in northern and western Davis Mountains; 510 ft. thick on Sheep Pasture M.; source or sources unknown. |
| Moore Tuff | 35.9 | Rhyolitic ash-flow tuff erupted from El Muerto caldera; up to 800 ft. thick in caldera, decreasing to 150 ft. 1½ miles outside caldera. |
| Frazier Canyon Formation | undated | Tuff and tuffaceous sandstone with lenses of pebble conglomerate; up to 330 ft. thick along Scenic Loop; source unknown. |
| Lower Tuff & Trachyte | undated | Rhyolitic porphyry and ash-flow tuff; reddish-brown; maximum thickness 240 ft. west of Crows Nest Hill; local unit; source unknown. |

Fig. 7.1: Formations seen along the Scenic Loop

# 7: DAVIS MOUNTAINS SCENIC LOOP

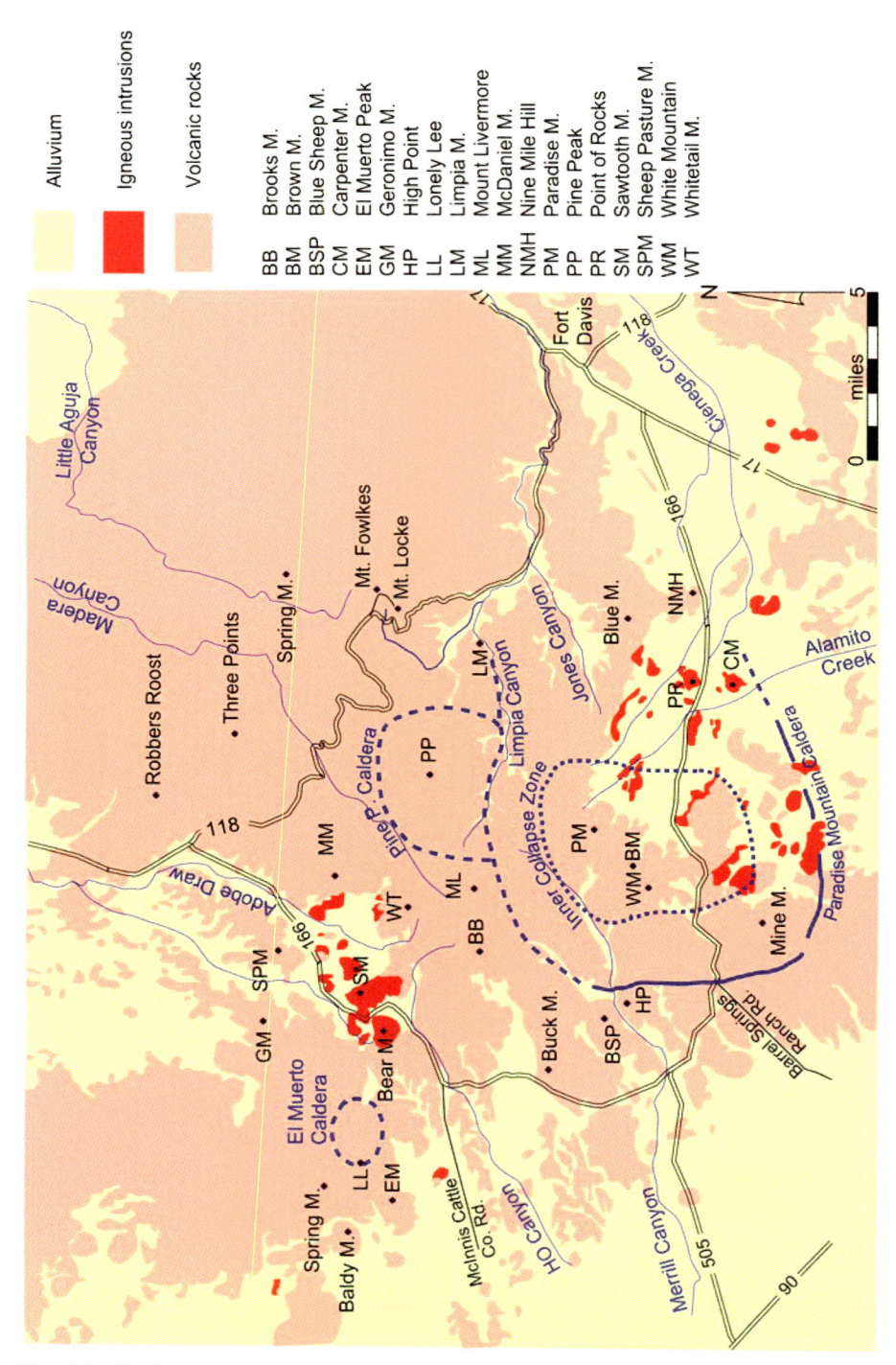

Fig. 7.2: Geology along the Scenic Loop

Fig. 7.3: Looking up Adobe Draw from the junction of Highways 118 and 166. Sheep Pasture Mountain is just to the right of center with Sawtooth Mountain looking over its shoulder. Geronimo Mountain is at the far right, like Sheep Pasture Mountain an outlier of Sheep Pasture rhyolite.

Fig. 7.4: The rounded syenite boulders of the Rockpile are good examples of spheroidal or onion-skin weathering. This type of weathering most often occurs when hot days and cool nights lead to daily expansion and contraction of jointed igneous rocks. Corners flake off at the joints, in time creating rounded forms.

## Stop 1: The High Mountains (3.8 miles from 118 Junction)

Just over a slight rise, the draw opens out as it approaches Sawtooth Mountain and a beautiful juniper-studded grassy meadow runs all the way up to the base of the mountains in the front left. McDaniel Mountain (7,235 feet) is the high point on the ridge at 9 o'clock with the rounded, heavily wooded top of Whitetail Mountain (7,484 feet) to its right. Two canyons which bracket the latter, Pine Canyon on the left and an unnamed one on the right, are the headwaters of Adobe Draw drainage.

## Stop 2: The Rockpile (5.5 miles from 118 Junction)

Leaving Adobe Draw the road enters a narrow pass between Sheep Pasture and Sawtooth Mountains and enters the drainage basin of Broke Tank Draw. The Rockpile is on the right, a very rocky sheer bluff with piles of rounded boulders (Fig. 7.4). At one time, it was a Department of Transportation roadside park and is shown as such on many maps, but the area, privately owned, has been closed off to public access for many years.

Both the bluff and the boulders are part of a syenite igneous intrusion, in which exfoliated and rounded knobs of pinkish brownish grayish syenite stick up abruptly from the plain. Exfoliation, or onion-skin weathering, is the process by which slabs of rock are stripped off large rock masses; eventually the masses become rounded. It results from the rock expanding as it heats up in the desert climate and contracting as it cools off. The syenite includes many feldspar phenocrysts up to a half inch long. They stand out on weathered surfaces, giving it a rough appearance. Many other syenite intrusions are found along the highway from this point on, the largest and most striking being Sawtooth Mountain at 10 o'clock from the Rockpile, 1,725 feet above the highway (Fig 7.5). Its extraordinary profile comes from jointing. Little Sawtooth Mountain (7,095 feet) on its left is a slump block, a block broken off the larger intrusion that came to rest on the slope below.

The bluff behind the Rockpile has similar jointing to Sawtooth Mountain, spaced irregularly. According to the Handbook of Texas Online, engineers of the State Department of Highways and Public Transportation in 1941 discovered the name "Kit Carson" and the date December 25, 1839 carved on a huge boulder here. In 1839, Kit Carson was a fur trapper who ranged widely over the western continent, and conceivably could have visited the Davis Mountains. However, he was illiterate in 1839 and only learned to sign his name as "C. Carson" later in life, so the inscription is almost certainly a fake. Another historical note is that the only grizzly bear found in Texas, a large and very old male, weighing about 1,100 pounds, was killed near Sawtooth Mountain in October 1890.

# 7: DAVIS MOUNTAINS SCENIC LOOP

Mount Livermore (8,381 feet) is behind Whitetail Mountain on the skyline at 3 o'clock joined by a rocky ridge on its right to Brooks Mountain (7,780 feet), the second highest of the Davis Mountains (Fig 7.6). The upper part of Brooks Mountain is a porphyritic trachyte, 1,000 feet thick, named the Brooks Mountain Formation by Anderson, and probably a volcanic dome developed late in the volcanic sequence. A quartz trachyte sill crops out on the lower slopes, overlying strata of the Wild Cherry and Barrel Springs Formations.

A half-mile beyond the Rockpile, a ranch road runs down Broke Tank Draw on the right past Geronimo Mountain (6,962 feet) at 3 o'clock, the last of the high Sheep Pasture hills seen on Highway 166. The draw runs north to join Adobe Draw near Flattop Mountain, 8 miles away.

The road continues slightly downhill bearing left between Sawtooth Mountain and the heavily wooded Broke Tank Draw and then begins climbing to the top of H O Hill at the York Ranch entrance. A large jointed syenite intrusion across the draw is conspicuous just before the entrance, part of the Sawtooth Mountain intrusion.

## Stop 3: York Ranch Entrance (8.1 miles from 118 Junction)

The summit of H O Hill is an excellent point to stop and view the terrain in front as well as the volcanic rocks under the Sawtooth Mountain intrusion. Large rounded boulders from the Sawtooth Mountain intrusion just before the entrance are underlain by quite finely foliated lava in a road cut, identified by Hoy as Moore Tuff (see page 109). This is most likely the base of the Sawtooth Mountain laccolith. A laccolith is a sill-like intrusion with a domed upper surface and a flat base.

Underlying the tuff is another syenite laccolith that extends to the Sawtooth Mountain foothills on the left and underlies Bear Mountain ahead. Boulders and outcrops of the laccolith at roadside going down H O Hill have very coarse grained syenite containing blocks of fine-grained rock. According to Nelson and Nelson, the latter developed first and was disrupted and incorporated in the former.

From the ranch entrance you can look down the hill to H O Canyon some 500 feet below; the rocky peaks of Mount Livermore are at 12 o'clock on the skyline with Brooks Mountain to their right. On the left along this section, a high ridge surmounted by a heavily jointed intrusion runs down from Sawtooth Mountain (Fig. 7.7). The round-topped Bear Mountain (7,256 feet) is on the right, 1,500 feet above the road. This summit is the highest point on Highway 166, elevation 6,212 feet, and on the drainage divide between streams that are in Pecos River drainage, such as Broke Tank Draw, and streams in Rio Grande drainage, such as the one

Fig. 7.5: Sawtooth Mountain (7,686 feet) from the Rockpile at Mile 5.5. The stunning profile is the result of jointing. Three sets are of mainly vertical jointing are present, at N 64° W, N 44° W, and N 45° E. The latter is perpendicular to the line of sight in this photograph.

Fig. 7.6: Mount Livermore (8,378 feet) on the left and Brooks Mountain (7,780 feet) from the Rockpile at Mile 5.5. These, the two highest summits in the Davis Mountains, are both thought to be surmounted by volcanic domes, created from very viscous magma in the final stages of volcanic activity.

Fig. 7.7: The intrusive ridge below Sawtooth Mountain from the bottom of H O Hill at Mile 7.2. The ridge is part of the Sawtooth Mountain intrusion of jointed pink syenite.

Fig 7.8: Mount Livermore on the left and the flat cap of Brooks Mountain on the right from a ranch entrance just beyond the picnic area at Mile 9.3.

running down H O Canyon below. The latter rises between Whitetail Mountain and Mount Livermore and disappears into the alluvium of Ryan Flat, but at one time would have drained into the Rio Grande.

The 600-foot difference in elevation between the two drainage basins is an interesting phenomenon because the rivers are at almost identical elevations at Pecos and Presidio, 2,550 feet, as you would expect. The controlling stream on the Rio Grande side is Alamito Creek, which runs through Marfa and then south to the river 6 miles downstream of Presidio. It has cut down through alluvium and volcaniclastic rocks its entire length. The Pecos drainage, in contrast, is cutting down through the much more resistant Adobe Canyon lava, hence the difference in elevations.

### Stop 4: Picnic Area below H O Hill (9.3 miles from 118 Junction)

The high ridge of Mount Livermore is at about 8:30 o'clock or so from the picnic area (Fig. 7.8), now with Brooks Mountain to its right. Just beyond, H O Canyon joins the road from the left. The hills on far side of the streambed are quite rounded and studded with trees, and two hundred yards ahead, at the ranch entrance on the left, the conical shaped hill at 9:00 and the light-colored hillside to its left are underlain by tightly folded marble and quartzite that continues along the canyon floor for about 1½ miles. Their light color stands out against the darker volcanic rocks; the area is known locally as the White Hills.

The rocks were identified tentatively as metamorphosed Finlay Limestone (the marble) and Cox Sandstone (the quartzite), of the Lower Cretaceous but their current elevation, 5,600 feet, is several thousand feet above the top of the Cretaceous limestone in nearby oil exploration wells which led some geologists to speculate that the limestone might have a fresh-water origin. Rainfall was plentiful during the volcanic era, and lakes formed on lava flows and ash-flow tuff beds in which limestone beds commonly developed.

However, Murley and Rohr found that the algae in the limestone were of the type found only in marine environments. They proposed that the block began as a roof pendant in the Sawtooth magma chamber, was metamorphosed by heat, lifted higher than now and dropped into place when the chamber subsided.

Continuing along H O Canyon, the road runs between lightly wooded hills on either side with the creek nearby on the left. At the mouth of the canyon, the road rounds a corner and the view opens out in front across Ryan Flat to Valentine and the Sierra Vieja ridge on the other side. The flat, a grassy meadowland with scattered yucca and cholla and occasional prickly pear, is underlain by the deep Valentine basin, one of a series of basins in the Salt Basin Rift described in Chapters 9 and 10.

## Stop 5: McInnis Cattle Company (12.1 miles from 118 Junction)

The McInnis Cattle Company entrance is on the right, in level pastureland with scattered trees and low cactus. The pointed hill at 3 o'clock on the horizon is El Muerto Peak (5,899 feet), named for the El Muerto Spring (also known as Dead Man's Hole) at its base (Fig. 7.9). Legend has it that a dead man was found near the spring when the Butterfield Stage began running stagecoaches between San Antonio and El Paso in 1854 or so. The stages used the spring as a watering point.

El Muerto Peak, Lonely Lee (6,394 feet) to its right and Spring Mountain (6,752 feet) behind and to the right of Lonely Lee, are trachyte laccoliths similar to the one at Bear Mountain. They were all intruded into Moore Tuff, an ash-flow tuff first mapped by Hoy and produced from the El Muerto caldera just north of Lonely Lee. The tuff is over 800 feet thick in the caldera, thinning to less than 160 feet 1½ miles out from the center.

A mile beyond McInnis Cattle Company entrance, the road crests a slight rise where you can look back at Bear and Sawtooth Mountains as shown in Figures 7.10 and 7.11. The rounded hills in the foreground are of Sheep Pasture rhyolite. Weathered Sheep Pasture outcrops are at road side as you approach San Antonio Pass.

## Stop 6: San Antonio Pass (14.7 miles from 118 Junction)

This minor uplift, 120 feet high, is a block of Sheep Pasture lava faulted against Merrill Formation quartz trachyte. The Merrill Formation is highly porphyritic with phenocrysts up to a half-inch long, similar to the Mount Locke Formation in appearance. One of the fault contacts is exposed in the left road cut.

The mesa, Buck Mountain (5,870 feet) on the left beyond the pass is capped by Barrel Springs welded tuff overlying Merrill Formation trachyte. Two and a half miles ahead, the hills on the right die away, opening up the view across Ryan Flat to the Sierra Vieja, 20 miles away at 3 o'clock. The rounded profile of Chinati Peak (7,730 feet) is at 12:30 and Capote Peak at 1 o'clock. These mountains are described in Chapter 10.

An interesting exposure is in the right road cut at Mile 16.7 where foliated ash-flow tuff overlies tuffaceous sediments that are possibly Wild Cherry tuffs. The ash-flow tuff is densely welded with some lava and pumice inclusions. The underlying tuff is oxidized and baked.

A half-mile ahead, the road crosses a draw and bears left to enter a long straight down towards the intersection with Ranch Road 505. Porphyritic trachytes of the Merrill Formation can be seen in several small road cuts and outcrops along this stretch.

Fig. 7.9: El Muerto Peak on the left from the entrance to McInnis Cattle Company ranch at Mile 12.1. Lonely Lee is on the right with Spring Mountain behind and to its right. All three mountains are laccoliths intruded into Moore Tuff. Fig. 10.15 shows Spring Mountain from the west.

Fig. 7.10: Bear Mountain on the left and Sawtooth Mountain from Mile 13.1. The ridge on the right of Sawtooth Mountain is shown close up in Fig. 7.6. The low dark hills in the foreground are of Sheep Pasture rhyolite.

feet above
sea level

- 8000
- 7000
- 6000
- 5000

Bear M. 7,256'
Highway 6,212'
Sawtooth M. 7,686'
Sawtooth M. ridge 7,110'

■ Syenite intrusion    ■ Sheep Pasture Formation
■ Quartz trachyte sill

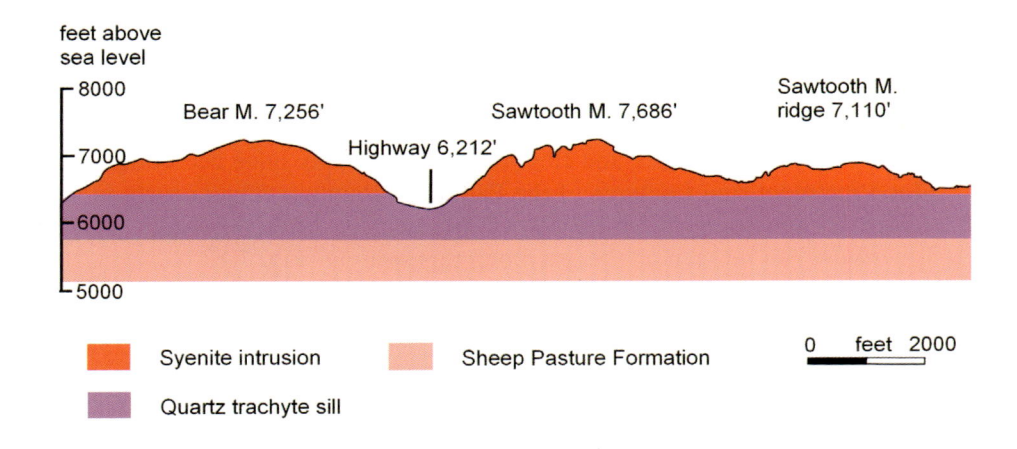

0  feet  2000

Fig. 7.11: Section across Bear and Sawtooth Mountains from the same viewpoint as Fig. 7.10.

Fig. 7.12: Another view of Brooks Mountain on the left and Mount Livermore from Mile 18.4. The massive bare rock summit of the latter is called Baldy Peak. The ragged Mount Livermore ridge is an eroded trachyte volcanic dome, which continues behind Brooks Mountain. The summit of Brooks Mountain is probably a volcanic dome, too.

The formation was named for Merrill Canyon ahead and is only found in this vicinity. Henderson concluded that it is either a series of trachyte flows or a volcanic dome and flow complex that may have been emplaced along incipient ring fractures of the Paradise Mountain caldera early in the caldera's development. Fine views of Brooks Mountain and Mount Livermore can be seen on the left along this straight (Fig. 7.12).

### Stop 8: CLE-VEL Ranch (19.3 miles from 118/166 Junction)

Blue Sheep Mountain (6,094 ft.) and High Point (6,230 ft.), both capped by Goat Canyon ash-flow tuff, straddle Merrill Canyon at 9 o'clock from the CLE-VEL Ranch entrance. Brooks Mountain and Mount Livermore are on the skyline to the left of Blue Sheep Mountain.

At Mile 19.5, Highway 505 (the Valentine Cutoff) joins the scenic loop from the right. Beyond it, the road winds between rounded grassy hills on either side.

### Stop 9: Lower Basalt Road Cut (21.5 miles from 118 Junction)

Three-quarters of a mile beyond the 505/166 junction, the road curves sharply left around a hill in a slight rise where brecciated basalt is exposed in a road cut. Broken surfaces have a light gray coating giving the road cut that color. The basalt is made up of several lava flows from an unknown source about 80 feet thick at maximum. It lies below the Merrill Formation in roughly the stratigraphic position of the Frazier Canyon Formation and may be one of that formation's mafic flows. At the eastern end of the road cut, the basalt is dropped into contact with Merrill Formation trachyte by a fault, marked by breccia and abundant secondary calcite mineralization at the contact.

A half-mile ahead, a low ridge of light colored rocks on the right is of the Lower Tuff and Trachyte. It underlies the Frazier Canyon Formation here and is about 240 feet thick. It consists of a lower welded dark reddish-brown, porphyritic ash-flow tuff, 20 to 40 feet thick, containing biotite phenocrysts towards the west, overlain by a porphyritic trachyte and then a welded ash-flow tuff which is partly vitrophyre. It generally forms the lower slopes of hills in the area with the trachyte forming small ledges of rounded boulders.

### Stop 10: Paradise Mountain Caldera (22.6 miles from 118 Junction)

A stream bed crosses the highway just before it bears to the right. It comes from Barrel Spring, about 1,500 feet to the left, where Henry Skillman set up his stagecoach watering point in 1850.

Immediately beyond the stream crossing, white silicified rhyolite crops out at the side of the road on the left and again a short distance ahead, followed by silicified basalt on the right (Fig. 7.13). These are outcrops of volcanic rock that has been altered by hot fluids under pressure, a process known as hydrothermal alteration.

Such fluids, predominantly steam, left over when other components of magma have crystallized, develop in the final stages of igneous activity. Steam if hot enough dissolves silica and other rock materials and as a gas can pass through the narrowest channels, dissolving some materials and depositing others. It often brings in valuable elements like silver, tin and copper and prospectors eagerly explore areas of hydrothermally altered rocks. Unfortunately, nothing of value has been found here.

According to Henderson, the source of the hydrothermal fluids was the Paradise Mountain caldera, which he discovered and named (Fig. 7.2). The western boundary of the caldera crosses the road at the Barrel Springs Ranch entrance, a half-mile ahead on the right. Its northern boundary may run along Limpia Creek; the location of the eastern boundary is uncertain.

Henry, Kunk and McIntosh established, on the basis of isotopic dating, that the caldera was the source of a set of tuffs which they call the tuffs of Wild Cherry, some of which had been identified by Henderson as belonging to the Barrel Springs Formation. These tuffs crop out in a northeast-southwest area 40 by 14 miles, and cross Highway 166 at the caldera boundary just ahead (Fig. 2.15). The formation is almost 1,600 feet thick on Paradise Mountain.

## Stop 11: Barrel Springs Ranch (23.2 miles from 118 Junction)

From the rise before the ranch entrance to about a mile beyond it, reddish colored volcanic rock, some of it silicified, can be seen in road cuts. Much of it has been broken up or brecciated into angular pieces, but whether this occurred during alteration or before is unknown. The silicified rocks are composed of various forms of quartz, which has replaced the original minerals in the breccia and filled in fractures and faults. According to Nelson and Nelson there are several stages of brecciation and resilicification present which suggest multiple episodes of silicification.

At the pullout a mile beyond the Barrel Springs Ranch entrance, a quite high roadside wall on the right exposes brecciated lavas for some distance. White Mountain (6,742 feet) is straight ahead, conspicuous by its cap of white silicified Wild Cherry tuff, 240 feet thick (Fig. 7.14). Just beyond the pullout, the road enters a steep-walled arroyo with a streambed on the left. Pinkish Frazier Canyon tuff is exposed in a road cut coming up the incline to Crows Nest Hill on the left.

Fig. 7.13: White silicified rhyolite crops out near the roadside at Mile 22.6. The lava is the result of hydrothermal alteration, alteration by hot fluids under pressure. The fluids dissolved the original constituents of the rock and replaced them by various forms of silica.

Fig. 7.14: White Mountain at sunset from Mile 25.2. The distinctive cap of white silicified Wild Cherry tuff can be seen for many miles to the south and east. Contrast the white material with the unaltered tuff on Brown Mountain (Fig. 7.15).

Fig. 7.15: Brown Mountain from Mile 28.0. The red lava cliff is of Wild Cherry tuff, 120 feet thick, overlain by 200 feet of Goat Canyon tuff. Note the white silicified Wild Cherry tuff to the left of the photograph.

Fig. 7.16: The bluff overlooking Skillman's Grove and the Bloys Camp Meeting Ground, taken from just beyond the Meeting Ground at Mile 29.2. The bluff has Wild Cherry tuff overlying a syenite intrusion.

At the top of the incline (Mile 25.9), an ash-flow tuff with an interbedded vitrophyre is exposed in road cuts on both sides of the highway. A small fault appears at the northeastern end of the outcrop.

The boundary of the inner collapse zone of the Paradise Mountain caldera crosses the highway just before Crows Nest Hill on the left. This zone of the caldera is the area of maximum hydrothermal alteration in the caldera. It measures 6 miles north to south and 4 miles wide.

## Stop 12: Crows Nest RV Park (27.8 miles from 118 Junction)

Mine Mountain (6,343 feet) is about a mile away at 3 o'clock from the Crow's Nest RV Park entrance. Rutile, an ore of the metal titanium, was discovered here in 1940. Titanium is used to harden steel and was much in demand just before World War II. The deposit was explored by surface trenches, pits and over 1,000 feet of underground workings but no economic reserves were found. The underground workings intersected a kaolin deposit in a steeply dipping shear zone in silicified tuff. The irregularly shaped deposit is about 400 feet long and up to 200 feet wide and has been drilled to a depth of 100 feet by sporadic investigations, the most recent in 1985.

Kaolin, a mixture of clay minerals, principally kaolinite, is used as a filler and in the manufacture of ceramics. Like the silicified lava, it is the product of hydrothermal alteration in the collapse zone of the Paradise Mountain caldera. The alteration is akin to that seen around hot springs where acidic water takes silica into solution leaving clay minerals behind and redepositing the silica elsewhere.

Henderson believed that a lake may have developed in the caldera as it collapsed and that water in the caldera seeped underground through faults and fissures, was heated by intrusions and circulated back up. It may have picked up some sulfur from the intrusions which would have created dilute sulfuric acid. The kaolin bodies would have formed where the hot water came to the surface. Titanium was probably leached from the tuff, a process that also required very acid water, and deposited as rutile in the kaolin zone.

Another kaolin deposit, called Medley's White Mountain Deposit, occurs just east of White Mountain (6,742 feet) on the other side of the highway (Fig. 7.14). Its existence has been known since the early twentieth century but it contains no rutile and no underground development has taken place. The deposit is tabular, 800 feet long, 200 feet wide and 50 feet thick and overlain by a silicified tuff cap. The red lava cliff at 9 o'clock on the skyline on Brown Mountain (6,983 feet) is of Wild Cherry tuff, 120 feet thick, overlain by 200 feet of Goat Canyon tuff (Fig. 7.15).

## Stop 13: Bloys Camp Meeting Ground (28.5 miles from 118 Junction)

Bloys Camp Meeting Ground on the right is clustered around the northern tip of a ridge (6,428 feet) consisting of a quartz syenite intrusion overlain by Wild Cherry strata (Fig. 7.16). Bloys Camp meetings were instituted in 1890 by William Bloys, a Presbyterian missionary in Fort Davis. Meetings began as a way of allowing ranching families and their employees to worship together. They take place for 5 days in August each year. The land where the meetings are held, Skillman's Grove, was bought by the Bloys Camp Meeting Association in 1902.

Skillman's Grove was named for Henry Skillman, military officer, scout, mail carrier, stage driver and Confederate spy, another of the extraordinary men who featured in nineteenth-century Texas history. Born in New Jersey, he first came to notice in the west as a captain in the Traders Battalion that fought in the Mexican War (1846-1848).

In 1849-50, he worked as a mail carrier between San Antonio and El Paso. In 1851, he was awarded the San Antonio-El Paso mail route by the Postmaster General, which he operated, not very successfully, until 1854. He set up four watering points in the Fort Davis area for his route, at Barilla Springs, Limpia Creek, Barrel Spring and El Muerto Spring. Later, he occasionally acted as driver and in one of his most famous exploits he drove the first Butterfield Overland Mail stagecoach in 1858 from Horsehead Crossing on the Pecos River to El Paso without dismounting, a feat that took 96 hours.

When the Civil War broke out and Union forces moved into El Paso, Skillman moved across the Rio Grande and spied for the Confederacy. He was tracked down and killed by Union troops near present-day Presidio in 1864.

Paradise Mountain (7,720 feet) is on the skyline at 9 o'clock from the straight beyond the encampment, on a long ridge that extends from the foothills of Mount Livermore nearly to Brown Mountain, capped by Goat Canyon rhyolitic ash-flow tuff (Fig. 7.18). The Haystacks are at 12:15 o'clock, Cathedral Mountain at 1 o'clock and Goat Mountain at 1:30 o'clock, all in the Alpine area.

Henderson mapped the eastern boundary of the inner collapse zone of the Paradise Mountain caldera crossing the highway at about Mile 30.3, a mile and a half beyond the Bloys encampment.

A number of syenite intrusions are exposed near the highway in the 4½ miles between Bloys encampment and the Point of Rocks roadside park including a medium-grained one at Mile 32.1 and a coarse-grained one 200 yards ahead.

Fig. 7.17: Point of Rocks intrusion from Mile 33.7. This side view shows the well-named intrusion tapering down to the roadside with many rounded boulders along its flank.

Fig. 7.18: A rocky ridge leads up to Paradise Mountain (7,720 feet) on the right skyline from Mile 32.0. The mountain is capped by Goat Canyon rhyolitic tuff overlying Wild Cherry tuff and Mount Locke porphyry. The two rocky syenite intrusions in front of the ridge are thought to be other outcrops of the intrusion that created Point of Rocks.

feet above sea level
— 7000
Blue Mountain 7,286'
— 6000
— road level 5,292'

■ Lava of Casket Mountain
■ Wild Cherry Tuffs
■ Tuffs and mafic lavas at stratigraphical position of Frazier Canyon Formation
■ Barrel Springs Formation

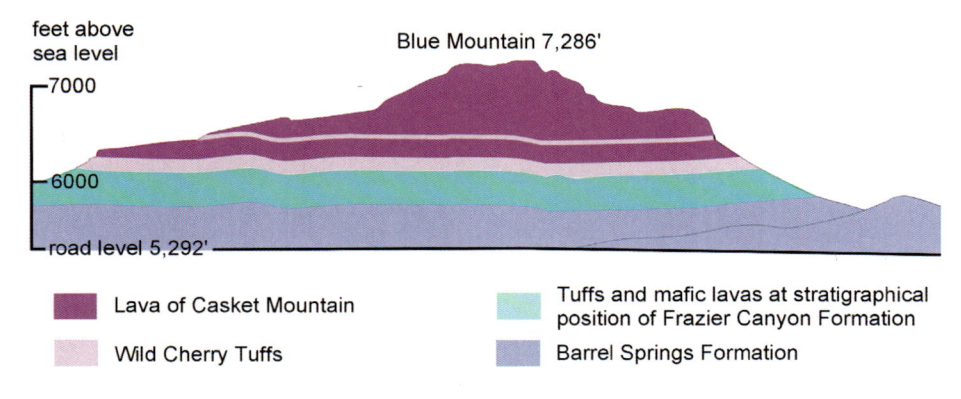

Fig. 7.19: Blue Mountain from Mile 33.5. The upper section is made up of at least 5 flows of the lavas of Casket Mountain, 1,080 feet thick, the lowest occurring below a 30 foot section of Wild Cherry tuff. The tuff, 160 feet thick in total, is underlain by tuffs and mafic lavas at the stratigraphical position of Frazier Canyon Formation. Barrel Springs strata make up the lower section and cap the hills to the left, including Nine Mile Hill

Fig. 7.20: Blue Mountain from Mile 37.0, on the east side of Nine Mile Hill, the opposite side from the section above. However, the stratigraphical succession is very similar.

## Stop 15: Point of Rocks Roadside Park (33.1 miles from 118 Junction)

The Point of Rocks Roadside Park has a syenite intrusion (5,920 feet) behind it (Fig. 7.17). This and the other intrusions in this area may be part of a single body (Fig. 7.2). They are most probably part of a dome that developed in the latter stages of the Paradise Mountain volcanic episode. Such domes, called resurgent domes by geologists, develop from magma rising up from below the magma chamber floor late in the eruptive cycle when the overlying magma has been removed by eruption. Carpenter Mountain (5,826 feet), south of the road, is another outcrop of the intrusion.

## Stop 16: Blue Mountain (35.1 miles from 118 Junction)

The massive presence of Blue Mountain (7,286 feet) rises 2,000 feet above the road ahead on the left. Thick lava beds crop out half way up and again at the summit (Figs. 7.19, 7.20). According to Henry, Kunk and McIntosh, the beds are five flows of the lavas of Casket Mountain, 1,080 feet thick. Four beds overlie 160 feet of Wild Cherry tuff, the lowest bed lying below 30 feet of the tuff. Tuffaceous volcaniclastics and mafic lava, equivalent in stratigraphical position to the Mount Locke Formation, are below the tuff, followed by Barrel Springs lava. The Barrel Spring ridge to the right of the mountain terminates with Nine Point Hill (5,630 feet), about 250 feet above the road.

## Stop 17: Blue Mountain Winery (36.3 miles from 118 Junction)

Blue Mountain Winery on the left is one of the oldest vinifera vineyards in Texas, first planted in 1977. The vineyard consists of seventeen acres at an elevation of 5,400 feet; the high elevation produces fine grapes. Four varieties are grown: Cabernet Sauvigon, Sauvignon Blanc, Chenin Blanc and Merlot. The operation does everything from vine to bottle (http://www.bluemountainwines.com). You can stop and tour this family-run winery.

Ahead, towards the junction with Texas 17, you can see across the northern wing of the Marfa plain to the Puertacitas Mountains (6,165 feet) at 3 o'clock. Several small Barrel Springs buttes are scattered on the plain.

The junction of Highways 166 and 17 is at Mile 42.4. Turn left on to Highway 17 to return to Fort Davis, 1.9 miles ahead. This part of the route is described in the next chapter.

# 8: Fort Davis – Alpine – Marfa – Fort Davis

## 69 miles

This loop takes in a wonderful variety of scenery including the Alpine basin, the Marfa plain, and two canyons with spectacular volcanic terrain. Musquiz Canyon, 5 miles from Fort Davis, descends through the lower Davis Mountains formations to the Alpine basin. Highway 90 between Alpine and Marfa goes through the chaotic jumble of a caldera's collapse zone.

Mileages from Fort Davis to Alpine are measured from the Jeff Davis County courthouse. The geology along the route is illustrated in Fig. 8.1 and the strata described in Fig. 8.4.

Driving south from the courthouse, bear left on to Highway 118 100 yards ahead. Dolores Mountain on the right is capped by a scraggly ash-flow sheet of the Barrel Springs Formation. The quarry at the tip of the mesa was the source of some of the pink stone used in the Limpia Hotel and Fort Davis State Bank buildings.

The sharp point of Mitre Peak is directly ahead on the horizon with the Haystacks (6,670 and 6,895 feet) at 1 o'clock. The Puertacitas Mountains (6,285 feet) are at 2:30 o'clock with the Mano Prieto Mountains (5,550 feet) in front.

Musquiz Dome is at 12:30 o'clock on the horizon, one of three domes along Musquiz Canyon created by intrusions into tuff horizons in the volcanic sequence. Two of them have been dated. They are about 32.8 million years old, 2½ million years younger than the youngest dated lavas in the Davis Mountains. Musquiz Creek, and the road that follows it, threads its way down the canyon between the domes (Fig. 8.3).

Musquiz Dome is the largest of the three domes (Fig. 8.2). In it, a rhyolite intrusion came in at the Huelster Formation level and thrust up a thick section of Star Mountain and Gomez Tuff strata into an elliptical dome.

8: FORT DAVIS – ALPINE – MARFA – FORT DAVIS

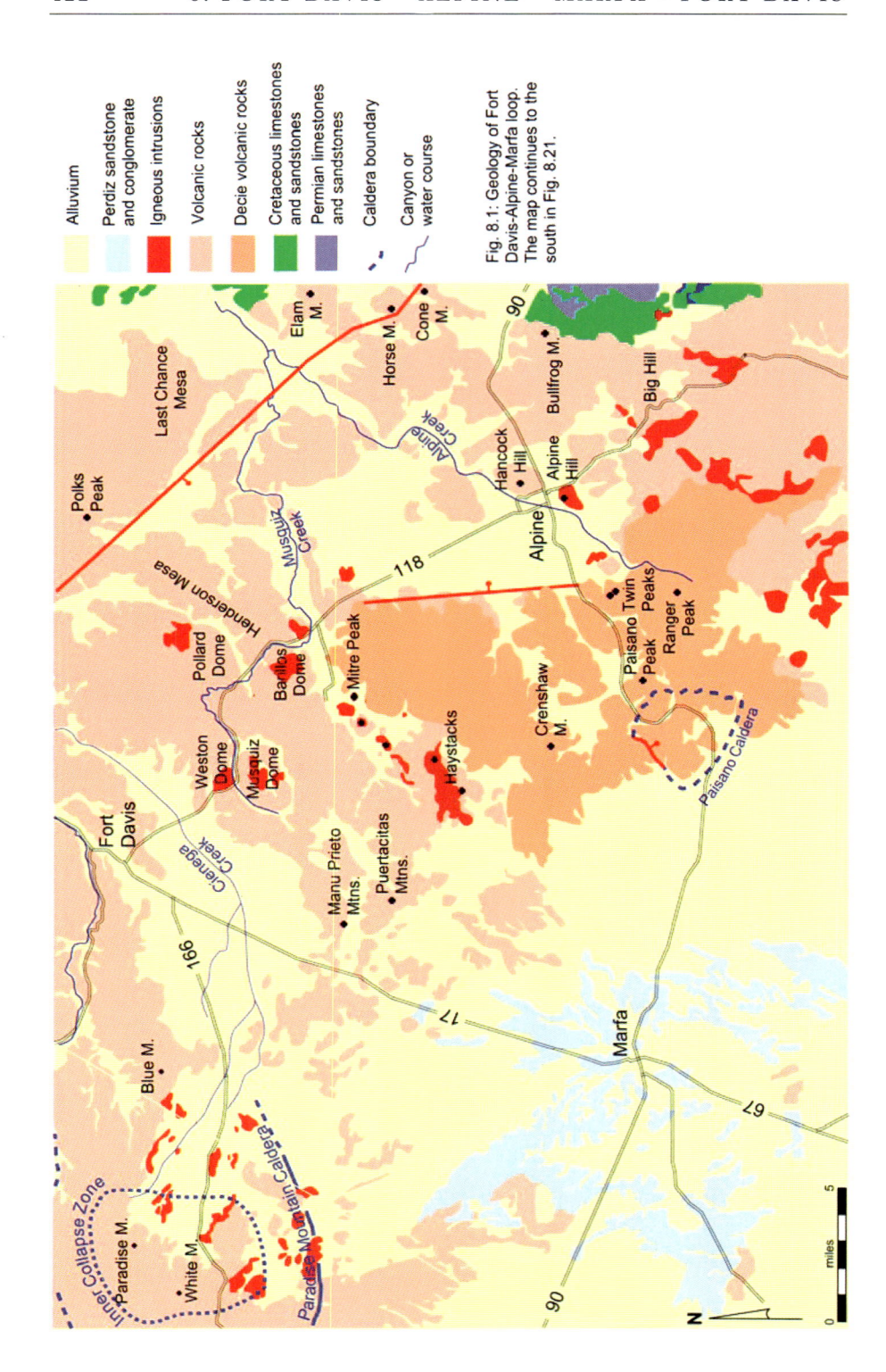

Fig. 8.1: Geology of Fort Davis-Alpine-Marfa loop. The map continues to the south in Fig. 8.21.

Fig. 8.2: The escarpment on the south side of Musqiz Dome from 5.8 miles south of Fort Davis on Highway 17. The prominent lava bed near the top of the escarpment is of the Sleeping Lion Formation with Barrel Springs strata above. The slopes below the Sleeping Lion cliffs are underlain by Frazier Canyon strata. Star Mountain Formation lavas are at the base of the visible section, dipping away from the dome at about 30 degrees.

Fig. 8.3: Musquiz Creek threads its way between the intrusions in Musquiz Canyon. The three domes are laccoliths intruded at either the Huester or Frazier Canyon tuff horizons. The Weston intrusion is a stock or a laccolith.

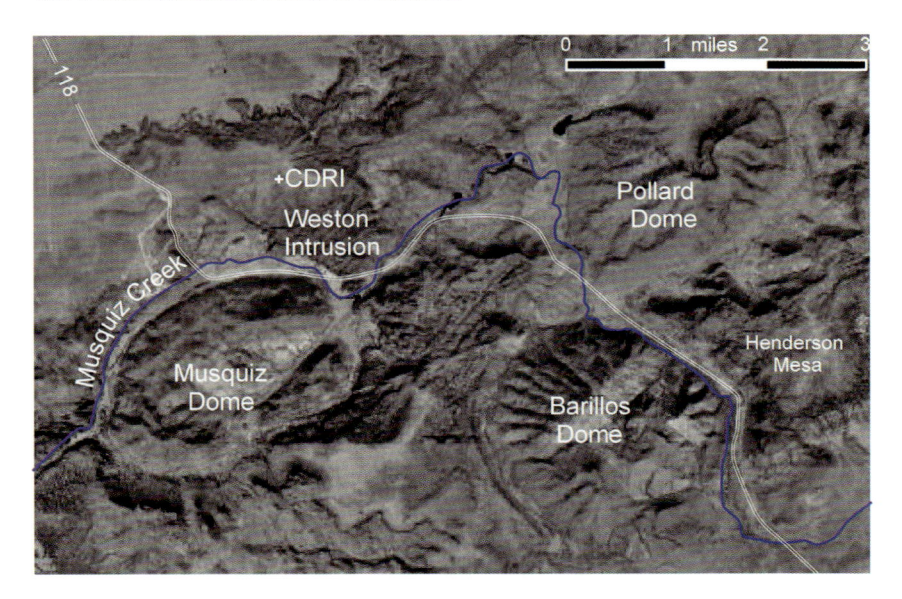

| Unit | Age m.y. | Description |
|---|---|---|
| Casket Mountain lavas | 35.3 | Up to 5 flows of porphyritic rhyolite capping ridges in southern, central and eastern Davis Mountains; probably erupted from widely spaced fissures. |
| Barrel Springs Formation | 35.6 | Upper unit: rhyolitic lava.<br>Lower unit: strongly rheomorphic ash-flow tuff or lava in central and eastern Davis Mountains; source presumably buried in central Mountains; up to 500 ft. thick. |
| Sleeping Lion Formation | 35.9 | Single rhyolite flow in central & southeast Davis Mtns., probably from source northwest of Fort Davis; 630 ft. thick at maximum. |
| Decie Formation | 36.3 | Rhyolite and quartz trachyte lavas with interbedded tuffs; source multiple fissures 10 miles west of Alpine; total thickness up to 3,000 ft. Five members identified:<br>*McIntyre Lava Member*: Quartz trachyte, trachyte, some rhyolite, minor agglomeratic tuff, conglomerate, and ash-flow tuff; up to 1,100 ft. thick.<br>*McIntyre Tuff Member*: Agglomeratic tuff and several small ash-flow sheets; 0-300 ft. thick, thinning towards Alpine.<br>*Morrow Lava Member*: Quartz trachyte, slightly porphyritic in places; thickness up to 750 ft.<br>*Morrow Tuff Member*: Yellow agglomeratic tuff inter-layered with two minor quartz trachyte ash-flow sheets; up to 300 ft. thick, thinning towards Alpine.<br>*Paisano Rhyolite Member*: Flows, domes and plugs of rhyolite with minor volcanic breccia, bedded tuff and poorly welded ash-flow tuff; up to 600 ft. thick. |
| Frazier Canyon & Cottonwood Springs Formations | undated | Tuff and tuffaceous sandstone with interbedded mafic lava flows; up to 1,200 ft. thick near Alpine, thinning northwards; occurs from 30 miles south of Alpine to 5 miles north of Fort Davis. |
| Limpia Formation | 36.5 | Quartz trachyte lavas overlying Gomez Tuff east & southeast of Fort Davis. |
| Gomez Tuff | 36.7 | Rhyolite ash-flow tuff in northern & eastern Davis Mountains; source is Buckhorn Caldera; 6 ft. thick in Musquiz Canyon. |
| Star Mountain Formation | 36.8 | Multiple rhyolite to quartz trachyte lava flows in eastern Davis Mountains; individual flows 200-600 ft. thick. |
| Crossen Trachyte |  | Two lava flows of porphyritic rhyolite to quartz trachyte from Musquiz Canyon to Elephant M.; grayish- to reddish-brown, weathers to rusty brown with pitted surface; maximum thickness 265 ft.; source probably multiple fissures buried by flows. |
| Huelster and Pruett Formations | 38.4 | Reworked tuffs; mafic lavas near base of Huelster Formation in N.E. Davis Mountains up to 500 ft. thick. |

Fig. 8.4: Strata seen along Fort Davis-Alpine-Marfa-Fort Davis loop.

The overlying lava must have been broken up by this action, breaching it as geologists would say, allowing erosion to develop along its axis and creating a deep valley which now cuts the dome in two and exposes the intrusion along the valley floor. Crossen lava is also found in the valley, its farthest west appearance. The Star Mountain lava remaining after erosion now forms escarpments on either side of the dome overlooking the valley.

It is likely that at the time of the intrusion, ground level would have been a good deal higher than it is now, certainly as high as the summit of Blue Mountain or the Haystacks, 7,000 feet or so, and perhaps higher. Therefore, drainage would have been well established before erosion brought the land surface down to the top of the dome at 5,950 feet.

The entrance to Powell Farms (Mile 2.7) is at the lowest point in the flat where Cienega Creek crosses the highway. This operation of Powell Farms grows cold-weather bedding plants for garden centers, nurseries, chain stores etc. Its growing season is August through October. Cienega Creek rises near Paradise Mountain and joins Limpia Creek in the Pecos River drainage basin (Fig. 8.1). Blue Mountain is on the horizon behind the Powell Farms buildings, standing alone 2,000 feet above the plain.

Looking back at 7 o'clock, a Barrel Springs flow has been eroded into a mesa, capped by Barrel Springs ash-flow tuff over Limpia Formation lava. The Sleeping Lion Formation, absent here, presumably lapped around but did not flow over the upper section of the Limpia Formation. A little farther to the right, however, another butte (5,140 feet), 260 feet above the plain, is capped by a small remnant of Sleeping Lion lava and Frazier Canyon tuff with Limpia Formation lava at plain level.

The highway climbs over the low shrub-covered ridge with the Star Mountain summit at 3:30 o'clock on the horizon and descends slightly to the Chihuahuan Desert Research Institute (CDRI) entrance. Boulders of Sleeping Lion rhyolite litter the roadside at the summit, underlain by Frazier Canyon tuff.

CDRI is a non-profit body set up in 1974 to study the Chihuahuan Desert region, which stretches from southern New Mexico and southeastern Arizona to just north of Mexico City, an area 800 miles long and 250 miles wide. CDRI operates the Chihuahuan Desert Nature Center and Botanical Gardens featuring a visitor's center, a 20-acre botanical garden, a large cacti and succulent collection, interpretive exhibits such as the Chihuahuan Desert Mining Heritage Exhibit, and several miles of hiking trails.

Fig. 8.5: This satellite photograph of the Alpine basin shows the main landmarks of the Alpine area with the faulting that created the basin superimposed. The deepest part of the basin is from Sunny Glen across to Hancock Hill where faulting has down thrown strata 1,250 feet on the right and 750 feet on the left.

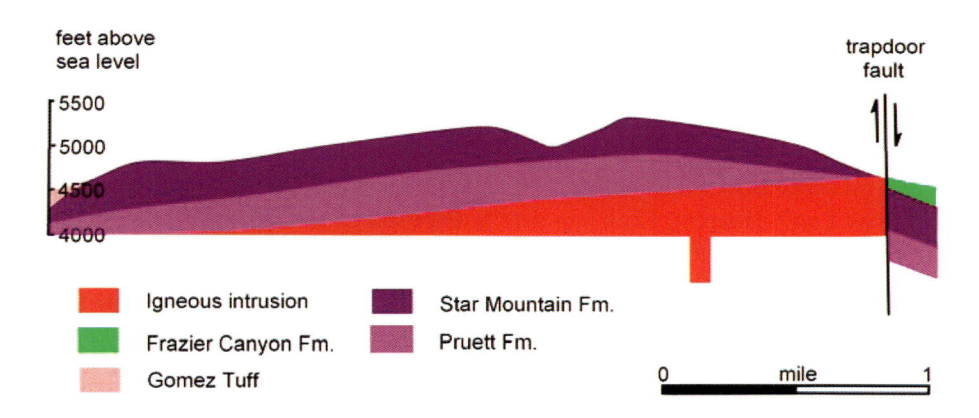

Fig. 8.6: Section across the Pollard Dome from the highway to a fault at the northwest corner of the structure. An intrusion came in at the Pruett Formation level, doming the overlying Star Mountain Formation and producing a fault that raised strata about 1,250 feet in the northwest corner of the dome. This type of intrusion is called a trapdoor laccolith, the fault being the trapdoor opening (adapted from Elkins).

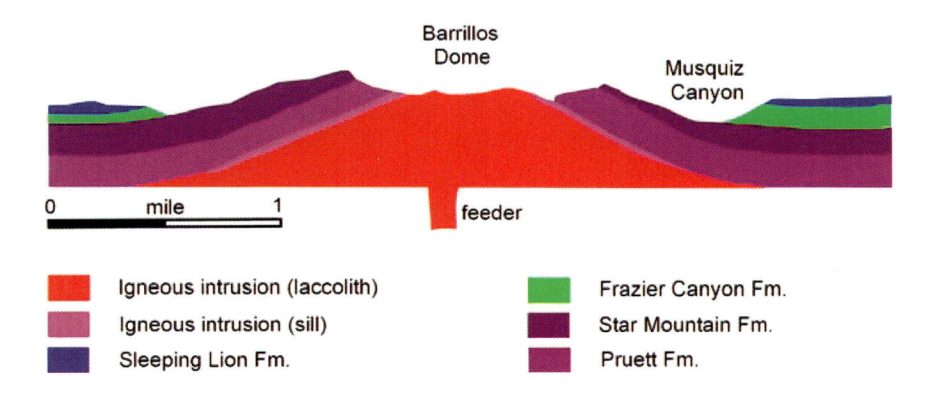

Fig. 8.7: Section across the Barillos Dome, a normal laccolith (adapted from Elkins).

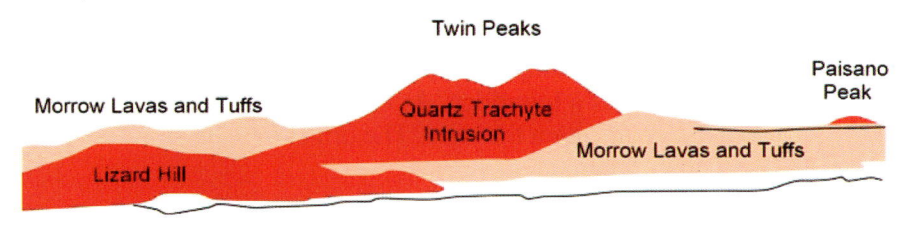

Fig. 8.8: Twin Peaks from the Big Bend Sportsman Club entrance 3 miles west of Alpine. The Lizard hill intrusion is separate from the Twin Peaks one.

Just beyond the CDRI entrance, the road bears right and drops 200 feet into the drainage of Musquiz Creek. The Weston intrusion on the left as you turn the corner is a stock or a laccolith intruded into Frazier Canyon tuff.

Half way down the incline, the gray-green faulted contact between the intrusion and Sleeping Lion rhyolite is exposed in the left road cut. At the bottom of the hill, a tongue-like projection of the intrusion penetrated and domed Frazier Canyon strata, which underlie Sleeping Lion lava (Fig. 8.10).

The road then bears left after crossing Musquiz Creek and for the next 1½ miles runs along the base of the Musquiz Dome where Gomez Tuff, overlying Star Mountain lava, dips down to the roadside. On the left, the cottonwood-lined creek runs along the base of the Weston dome on the far side of an attractive meadow. Note the remnant of Sleeping Lion lava on top of the light-colored smooth slopes of the intrusion. Half a mile ahead, a small canyon comes down along the edge of the intrusion to the creek.

An historical marker at Mile 6.2 identifies the homestead of the eponymous Manuel Musquiz, who moved here in 1854, fleeing unsettled conditions in Mexico. Life became difficult for the Musquiz family when Union troops withdrew from Fort Davis at the beginning of the Civil War. The family no longer had protection from marauding Indian bands and had to return to Mexico. According to the Handbook of Texas, an old grave was dug up at the old ranch buildings in 1907. Beneath the rotted coffin and disintegrating bones was an empty depression in the earth, and the footprints of a man, a woman, and a burro leading away from the grave. Area residents speculated that Musquiz family members had returned from Mexico to dig up some long-buried treasure, but the story was never confirmed.

At the pullouts at Mile 7.0, thick rounded columns of Sleeping Lion Formation form the mesa caprock on the left and produce the rounded boulder-strewn slopes below.

From Mile 7.7 to 8.9, road cuts display Limpia Formation trachytes overlain by bedded tuff of the Frazier Canyon Formation.

At Mile 9.4, the creek swings to the left and navigates a narrow gorge while the highway crosses a low ridge. At the summit of the ridge, Gomez Tuff is exposed in a road cut on the right. The tuff is only about 6 feet thick here near the southern limit of its extent.

The Pollard Dome is at left front. In it, Star Mountain Formation has been uplifted by an intrusion only exposed in a fault escarpment in the northeastern corner. The fault suggests that the intrusion is a trapdoor laccolith, with the fault being the open door (Fig. 8.6).

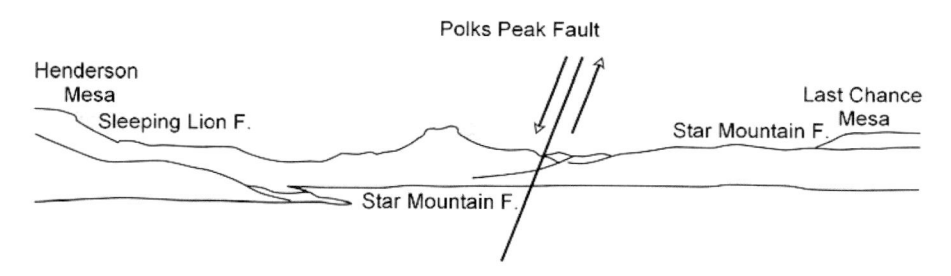

Fig. 8.9: Polks Peak viewed from the McElroy Ranch entrance at Mile 19.1. The Polks Peak fault brings down the Star Mountain Formation from 5,500 feet elevation on Last Chance Mesa on the right to 4,560 feet at the base of Henderson Mesa on the left. The Sleeping Lion Formation is absent to the right of this fault.

Traveling down to hill, the road cut on the left is in Star Mountain rhyolite at the corner of the Pollard Dome. Just beyond the entrance to the CF Ranch, Kokernot Creek crosses the highway to join Musquiz Creek and a canyon forms ahead along the edge of the Barillos Dome, which is, according to Elkins, a rhyolite intrusion into Pruett tuff domed by a later trachyte laccolithic. Star Mountain rhyolite dips off the intrusion to the north.

The historical marker at Mile 10.5 is for the first rural school west of the Pecos River.

## Stop 1: Pullout at the Barillos Dome (11.4 miles from Fort Davis)

The smooth light brown slopes of the intrusion are at 3 o'clock from this pullout. High jointed escarpments of Sleeping Lion lava are on the horizon behind and to the right of the dome, underlain by Frazier Canyon tuffs.

Henderson Mesa above the road on the left is capped by Star Mountain lava. Bedded tuff and mafic lava of the Frazier Canyon Formation are exposed in road cuts just beyond the pullout.

Mitre Peak (6,190 feet) and Antelope Peak (5,848 feet) are at 3 o'clock from the next pullout at Mile 12.8. The rhyolite bodies, and two others hidden behind them, have been intruded along a fault zone. Mitre Peak's rough surface comes from erosion along jointing. A tuff of unknown origin has created the thin ledge about a third of the way up the peak. A dike perhaps related to the intrusion forms a ridge in the foreground just to its right.

At Mile 13.2, a quartz trachyte sill intrudes Frazier Canyon Formation in a road cut on the left.

Fig. 8.10: An irregular tongue of the Weston intrusion penetrates Frazier Canyon tuff near Musquiz Creek at Mile 4.7. The intrusion is on the right overlain by tuff baked red by the heat of the intrusion. Mafic lava in the Frazier Canyon Formation is on the upper left.

Fig. 8.11: The Barillos Dome from the top of the rise at Mile 13.8. The intrusion at the center of the dome is the light brown mass to the right of the light gray Pruett tuff. Sleeping Lion lava forms cuestas to the left and behind the intrusion. Fig. 8.7 explains the structure.

Fig. 8.12: Looking north along the Alpine basin to Polks Peak from Mile 19.1. Fig. 8.9 explains the structure.

Fig. 8.13: The opening of Sunny Glen in the Decie cliff at Mile 19.1. The break about halfway down the cliff is between the Morrow and McIntyre lavas. The light colored patches below the cliff behind the left steer are from interbedded tuffs in the Cottonwood Springs Formation. The top of this formation on the escarpment is about 5,100 feet compared to 4,350 feet in a well drilled near the highway.

Musquiz Creek is joined by Barillos Creek from the right and crosses the highway at the low point in this loop (Mile 13.5, 600 feet below Fort Davis) and runs north to the Pecos River. The quite pronounced creek valley, 90 feet below the ridge in front, is in line with the fault along which Mitre Peak and the other peaks were intruded and is probably the result of erosion along the fault. Faults aid erosion because rocks tend to get broken up along them, and broken-up rocks are more easily eroded than unbroken rocks.

The photograph of the Barillos Dome (Fig. 8.11) is taken from the slope ahead. The light-colored bedded tuff on the left of the intrusion is from the Pruett Formation and are overlain by strata of the Star Mountain, Frazier Canyon and Sleeping Lion Formations. The latter is the mesa caprock at 6:30 to 9 o'clock. Note the outcrop of Frazier Canyon Formation above the valley at 7:00 with the dark band of mafic flows in the otherwise light gray outcrop.

The taller of the two Haystacks (6,895 feet), the highest mountain in the Alpine area, is on the skyline to the right of Mitre Peak. The other Haystack is hidden behind it. You can see both from Highway 90 near Marfa (page 141).

## Stop 3: McElroy Ranch Entrance (19.1 miles from Fort Davis

Up the rise from the Musquiz Creek valley, the Alpine basin lies in front, about 22 miles long and 7 miles wide. The best view of the basin as a whole is from the entrance of this residential development where the cliffs on the right are in full view (Fig. 8.13).

Looking back, you can see the basin running north to Polks Peak (5,304 feet), the hat-like butte 14 miles away at 7 o'clock (Fig. 8.12). A fault runs northwest-southeast just to the right of the peak and displaces strata about 900 feet down to the left. Thus the Star Mountain lava that caps Last Chance Mesa on the right of the peak runs along the base of Henderson Mesa on the left (Fig. 8.9). Sleeping Lion lava caps Henderson Mesa with Frazier Canyon tuff occasionally showing on the escarpment below. Polks Peak, too, is capped by a small Sleeping Lion outlier.

A rolling alluvial plain stretches across to low rounded hills of a Crossen ridge (4,550 feet) on the left. At 3:30 o'clock, the South Orient railroad follows a canyon 350 feet deep cut through the ridge by Alpine Creek.

The hills on the horizon are on the upthrown side of the Polks Peak fault. The highest is Elam Mountain (4,940 feet) at about 3 o'clock with Horse Mountain (4,890 feet) to its right.

Gorski measured the displacement of the Polks Peak fault where it crosses the Orient line at 500 feet up to the northeast. He wrote that it could be traced to the McCutcheon fault that crosses Highway 17 near Wild Rose Pass (see page 60). To the southeast, the fault continues from Horse Mountain through the gap between the Glass Mountains and Bird Mountain at 11 o'clock and lines up with a prominent break running across the Marathon Basin. The fault persists for an enormous distance for one with such a small displacement, and must be the surface embodiment of a major fault in the Proterozoic crust far below.

As you look south across the basin, the limestone slopes of the Glass Mountains dominate the skyline from 10 to 11 o'clock about 20 miles away. They are faced mainly with Permian Capitan Limestone. Bird Mountain, whose rounded top is on the skyline at 11 o'clock, is at the north end of the Del Norte Mountains, part of a Laramide-age mountain range that continues south for 70 miles to the Rio Grande and then for a further 60 miles into Mexico.

Hancock Hill (4,925 feet), with Sul Ross State University on its flanks at 11:30 o'clock, is a block of Crossen lava about 500 feet above the plain. Mount Ord (6,803 feet), whose triangular shape is on the horizon behind it, is the highest point in the Del Norte Mountains. The right slope on Mount Ord is faced with Crossen lava. The mountain is on the left wing of a syncline that formed since volcanic activity ended.

Barillos Creek marks the approximate boundary between rocks that came from volcanic centers to the north and rocks of the Decie Formation from the Paisano volcano. Lava flows from the latter cap the 1,000-foot escarpment on the western boundary of the Alpine basin. They consist of the upper McIntyre lava and the lower Morrow lava, separated by a thin layer of tuffaceous material that creates a distinct break in the cliffs. The Morrow and McIntyre tuff members do not occur along the escarpment (Fig. 8.4). The McIntyre unit consists of three flows, the Morrow two flows, with only slight differences between them. Breccia zones 10 to 15 feet thick are common between the flows. Both units exhibit well-developed columnar jointing. Canyons have developed at intervals along the escarpment, the deepest being Sunny Glen at 3 o'clock where the two lava units are 600 feet thick. They thin to about 50 feet at Mitre Peak.

The Decie Formation is underlain by the Cottonwood Springs Formation (Fig. 2.13), basalts with interbedded tuff that are exposed along the lower part of the escarpment (Fig. 8.13) and are found in numerous water wells drilled in the Alpine basin. The formation is similar in composition and age to the Frazier Canyon Formation. Below it is Crossen Trachyte, a rhyolitic lava flow very similar to the Star Mountain rhyolite in composition and age.

Fig. 8.14: Ranger Peak on the left and Twin Peaks from Mile 2.9. The mesa to the right of Twin Peaks is capped by a Morrow lava flow.

Fig. 8.15: This dike across the highway from the picnic area at Mile 5.3 is one of perhaps a thousand found in the Paisano volcano. They began as fissures that carried magma up to the surface. Magma in the fissures solidified at the end of volcanism, creating the dikes. McIntyre lava caps the escarpment on the skyline.

The Alpine basin appears to be a graben, at least towards its southern end where faults on either side lower strata into the basin between Sunny Glen and Hancock Hill (Fig. 8.5). At Sunny Glen the fault appears to have had a displacement of 750 feet (Fig. 8.13). Across the basin, the fault at Hancock Hill has a displacement of about 1,250 feet. The satellite photograph suggests that another minor fault hidden in the alluvium runs along this side of the ridges in the middle distance (shown by the dashed fault line on Fig. 8.5), displacing strata an unknown amount towards the highway.

## Alpine

The highway enters Alpine at Mile 22.3. The city began life around Kokernot Springs on the wagon trail between Chihuahua City in Mexico and San Antonio. Settlers arrived when the Galveston, Harrisburg and San Antonio Railway came through in 1882 and made Alpine a watering stop. The town quickly became a shipping point and supply center for ranching and was made county seat in 1884 when Brewster County was split off from Presidio County.

In 1921, a summer normal school, now Sul Ross State University, began classes leading to an increase in the population from 921 in 1920 to 3,495 in 1930. Since then, the town has grown slowly; in the 2000 census, the population was 5,786. Today, Sul Ross is the largest employer, and ranching and tourism are important industries.

The intersection of TX 118 and Highway 90 is at Mile 23.5. Turn right on to Highway 90 towards Marfa and reset your odometer.

On leaving Alpine, the high lava cliffs of the Decie Formation are to the left and ahead. The road climbs a sloping erosion surface or pediment as it approaches the cliffs. Lizard Mountain is the rough quartz trachyte intrusion at 9 o'clock from Mile 2.4, with a trachyte dome that was uplifted by an intrusion on its right. The intrusions on the skyline are Ranger Peak (6,246 feet) 2½ miles away at 10 o'clock, and the double-headed quartz trachyte Twin Peaks (6,133 feet and 6,112 feet) at 11 o'clock (Fig. 8.14).

Two openings separated by a ridge 750 feet high carry the highway and the railroad into the mountains. Horizontal quartz trachyte flows of Morrow lava cap the high ground on either side, underlain by Morrow tuffs with Cottonwood Springs basalt cropping out at the base of the escarpments. At Mile 3.8, you can look up the fault scarp running north along the skyline at 1 o'clock. The fault, which drops strata 750 feet into the Alpine basin at Sunny Glen, is intermittently exposed between there and the railway line.

At Mile 4.8, now in the canyon, the ridge above on the right is crowned by columns of light brownish-gray Morrow lava up to 300 feet thick. The large boulder at the roadside came from the cave above. Occasional outcrops of light gray tuff can be seen on the ridge flanks.

Just beyond the boulder, a much jointed Morrow quartz trachyte lava flow crops out in a road cut on the right with a 10-ft thick dike of inclusion-filled tuff near its east end. When eruptions at the Paisano volcano ended around 36 Ma, lava in the fissures solidified into dikes, hundreds of which crop out in the volcano area today.

## Stop 4: Picnic Area (5.6 miles from Alpine)

A ridge across the highway from the picnic area is surmounted by the ragged crest of one of the Paisano volcanic feeder dikes, which continues parallel to the road for the next half mile (Fig. 8.15). Others can be seen behind the picnic area.

The high walls around the area are capped by McIntyre lava underlain by Morrow lavas and tuff. At Mile 5.8, a block of yellowish-gray bedded Morrow tuff is faulted down against pale pink to dark red Paisano rhyolite. In places, the rhyolite is a very pretty spotted rock, light bluish gray on fresh surfaces, which was given the name *paisanite* by one of the early geologists to visit the area.

In front, the elegant rounded silhouette of Paisano Peak (6,085 feet) rises 1,050 feet above the road, a nepheline syenite plug intruded into the Paisano Rhyolite after volcanic activity had ended.

At the top of the rise at Mile 6.3, outcrops of Morrow welded ash-flow tuff, bedded in places, include boulders several feet across. Its color on fresh surfaces varies from pink to dark green and yellowish-gray. In the left road cut, the tuff includes a circular trachyte inclusion, 3 feet in diameter. No trachyte erupted in the volcano before the Morrow tuff, so this and other trachyte inclusions must have been brought up from the Earth's crust below the Decie strata.

At the top of the next incline, Mile 7.2, high cuts bracketing the road expose very shattered spotted paisanite (Fig. 8.16). The left road cut runs along the edge of a quartz trachyte dike cutting the rhyolite. Striations in the gray wall show that the dike flowed at roughly 60 degrees up to the east (Fig. 8.17).

## Stop 5: Gage-Holland Ranch (7.5 miles from Alpine)

The railway line comes into view on the right at the Gage Holland ranch entrance. Crenshaw Mountain (5,965 feet) is at 2 o'clock across upper Sunny Glen, capped by McIntyre lava, as are the uplands to its left.

Just after the ranch entrance, a low road cut shows welded blue gray Morrow tuff grading into poorly welded cream-colored tuff at the west end of the cut.

## Stop 6: Paisano Pass Caldera (9.0 miles from Alpine)

The highway crosses on to a complex of broken-up lava and tuff, called collapse terrain on the 1:250,000 scale geological map, opposite a light-gray railroad bridge on the right. The complex, 3 miles in diameter, was first identified as a possible caldera by Don Parker, who mapped it as part of his doctoral dissertation at the University of Texas in 1976. He had identified the Buckhorn Caldera in his M.A. thesis 6 years earlier.

The caldera formed after eruption of the McIntyre tuff and before eruption of McIntyre lava; strata older than the latter are very broken up, strata younger only slightly so. Displacement into the caldera along its perimeter varied widely, from 1,000 feet west of Paisano Peak to negligible amounts in other places. Road cuts along the next 4 miles show how the caldera collapse created jumbled groups of rocks caught up in mud flows. The climate during the volcanic era in west Texas was wet and warm, leading to plentiful mud.

The caldera boundary runs along the low ground between the railway line and the high hills on the skyline at about 12:30 o'clock and crosses the road again near the South Orient Railroad line 6 miles ahead. It then runs approximately parallel to the highway along the high hills behind the Paisano Baptist Encampment and returns to this point passing about a half-mile west of Paisano Peak, which is outside the caldera.

The bluff ahead on the left is in the caldera. The rounded hill at 3 o'clock is a rhyolite dome outside the caldera and the smooth yellow hills to its right also appear to be rhyolite domes.

In the road cut from Mile 9.5-9.8, the left wall shows the tremendous variety in the caldera. Much of the wall overhangs the ground below and is best viewed from the right side of the road. It begins on the left with a fossil mudflow containing a red brown tuff block oriented at 45 degrees to the left (Fig. 8.19), next a patch of welded volcanic breccia in red brown tuff, followed by a large block of volcanic breccia between two sections of poorly welded yellow gray tuff 10 and 3 feet wide, then a section of very poorly welded tuff with red-brown blocks of volcanic breccia, and at the west end, red-brown welded tuff. The jumbled nature of the road cut and its randomly oriented blocks suggest that it resulted from a landslide into the caldera.

Another similar road cut is at Mile 10.0.

Fig. 8.16: Heavily shattered Paisano Rhyolite, the lowest member of the Decie Formation, in the right road cut at Mile 7.2. The yellow coloration is superficial, deposited by fluids coursing through the fractures. On fresh surfaces, the rock is light gray with blue clumps of arfvedsonite, a mineral of the amphibole group. Osann, an early visitor to the area, called this rock paisanite.

Fig. 8.17: The wall of a quartz trachyte dike is exposed in the left road cut at Mile 7.2. Lineations in the dike, created as it flowed around irregularities in the rhyolite wall rock, show that the magma flowed up from right to left at about 60 degrees.

Fig. 8.18: Looking across to Sunny Glen from Mile 9.5. Crenshaw Mountain is on the left. The cliff above the Sunny Glen opening from the Alpine basin is on the far right.

Fig. 8.19: A large block of brown Morrow tuff caught up in a mudslide is exposed in the left road cut at Mile 9.5. This long road cut is a fine example of the chaotic nature of collapse in active volcanic terrain.

Just beyond the Paisano Baptist Encampment entrance at Mile 10.7, the thick dike exposed in a road cut and creating a ridge on the right is 3 miles long, one of the largest mapped in the volcano. The 500-foot high walls at 9 o'clock behind the encampment are on the western edge of the caldera where the collapse was around 1,000 feet.

A little ahead, Mile 11.2, reddish tuff containing chaotic accumulations of coarse angular volcanic rocks overlies light gray tuff in a road cut on the left.

At Mile 11.4, road cuts on both sides of the road are similar to the one at Mile 9.5 in that lava autobreccia, rounded tuff boulders, angular welded tuff blocks and foliated lava can all be seen in a mudslide matrix.

Finally, as the highway approaches the summit of the pass, patches of dark blue-gray unidentified rock can be seen in mudflow matrices on both sides of the road.

## Stop 7: Paisano Pass (13.2 miles from Alpine)

An historical plaque marks the summit of the pass (5,074 feet). When the Galveston, Harrisburg and San Antonio Railroad built its track in 1882, the pass was said to be the highest point on the line between New Orleans and Portland, Oregon. It is also claimed to be the highest point on Highway 90 between Florida and California. The basalt-capped hill at 9 o'clock is in the caldera.

At Mile 13.6, the highway crosses the South Orient Railroad. This line, originally the Kansas City, Mexico and Orient Railway, was built to run the 1,600 miles between Kansas City, Missouri, and Topolobampo, Mexico. It was completed to Presidio in 1930, sharing tracks with what is now the Union Pacific line from Alpine to this point. In 2000, the Texas Department of Transportation purchased the line from its previous owners and leased it for 40 years to the Mexican-owned company Texas Pacifico Transportation Ltd., which is bringing the track into service.

Just beyond the railway bridge, the road crosses the western boundary of the caldera on to the Marfa plain and one of the great vistas in the Big Bend opens up. The plain east and south of Marfa is underlain by lavas and volcaniclastic rocks and by large expanses of the coarse Perdiz Conglomerate (Fig. 8.21). Outcrops of this rock are widespread around Marfa and along Highway 67 to Presidio. It is a fanglomerate, i.e. a cemented alluvial fan. Its source is most probably the Chinati Mountains, which at one time must have been much larger than now. Possibly some of its source material around Marfa came from the southern Davis Mountains.

| Cathedral Mountain | Cienega Mountain | Goat Mountain |

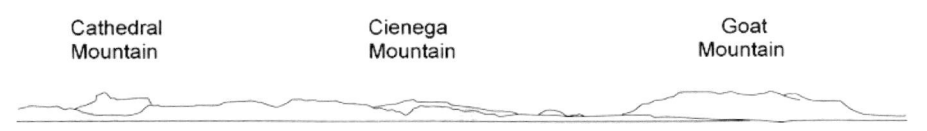

Fig. 8.20: The skyline at 6 o'clock from the Marfa Lights Viewing Center.

The Chinati Mountains border the plain on the southwest. Their highest point, Chinati Peak (7,728 feet), is the rounded mountain on the horizon, 44 miles to the southwest at 10:30 o'clock, a complex of large intrusions and lava flows erupted from the Chinati volcanic center and very similar in composition to those of the Paisano volcano. The large rhyolite lava dome of the Cienega Mountains (5,223 feet), 4 miles across and 1,400 feet high, is low on the horizon to the left of Chinati Peak. The volcanic Bofecillos Mountains, over the horizon 50 miles away, form the southern boundary of the plain.

## Stop 8: The Marfa Light Viewing Center (17.2 miles from Alpine)

The Marfa Lights Viewing Center is a good place to stop and look over the area. The twin Haystack trachyte intrusions (6,895 and 6,670 feet) are on the horizon a little after 3 o'clock. Their age is unknown; a nearby intrusion was dated at 34.6 Ma, about 700,000 years after the main phase of Paisano volcanic activity had ended.

Nearer the road, the Black Peaks, several small dark gray-brown intrusions, stick up noticeably out of the surrounding Morrow lava at 4:30 o'clock. A sample was dated at 36.9 Ma, near the beginning of volcanic activity. The Puertacitas Mountains (6,285 feet), on the skyline at 2:30 o'clock, consist of three small brown-tinted intrusions in basalt.

Looking back towards Alpine, Paisano Peak is at 5:30 o'clock and Cathedral Mountain (6,800 feet), at 7 o'clock. The spire on the left of Cathedral Mountain is a block faulted up by an intrusion below. The mountain is capped by a thick basalt flow overlying volcaniclastic strata from the Chinati volcanic center. The flat-topped bench to its left is capped by Mitchell Mesa Welded Tuff. The tuff, a widespread, thin and very durable rhyolite produced from the Chinati Mountains caldera at about 33 Ma, is found as far as 50 miles northeast and 50 miles south of the caldera.

Cienega Mountain (6,562 feet) just to the right of Cathedral Mountain is a large intrusion that arched overlying strata when it was emplaced so that Cretaceous limestones now crop out on its west flank facing us.

142     8: FORT DAVIS – ALPINE – MARFA – FORT DAVIS

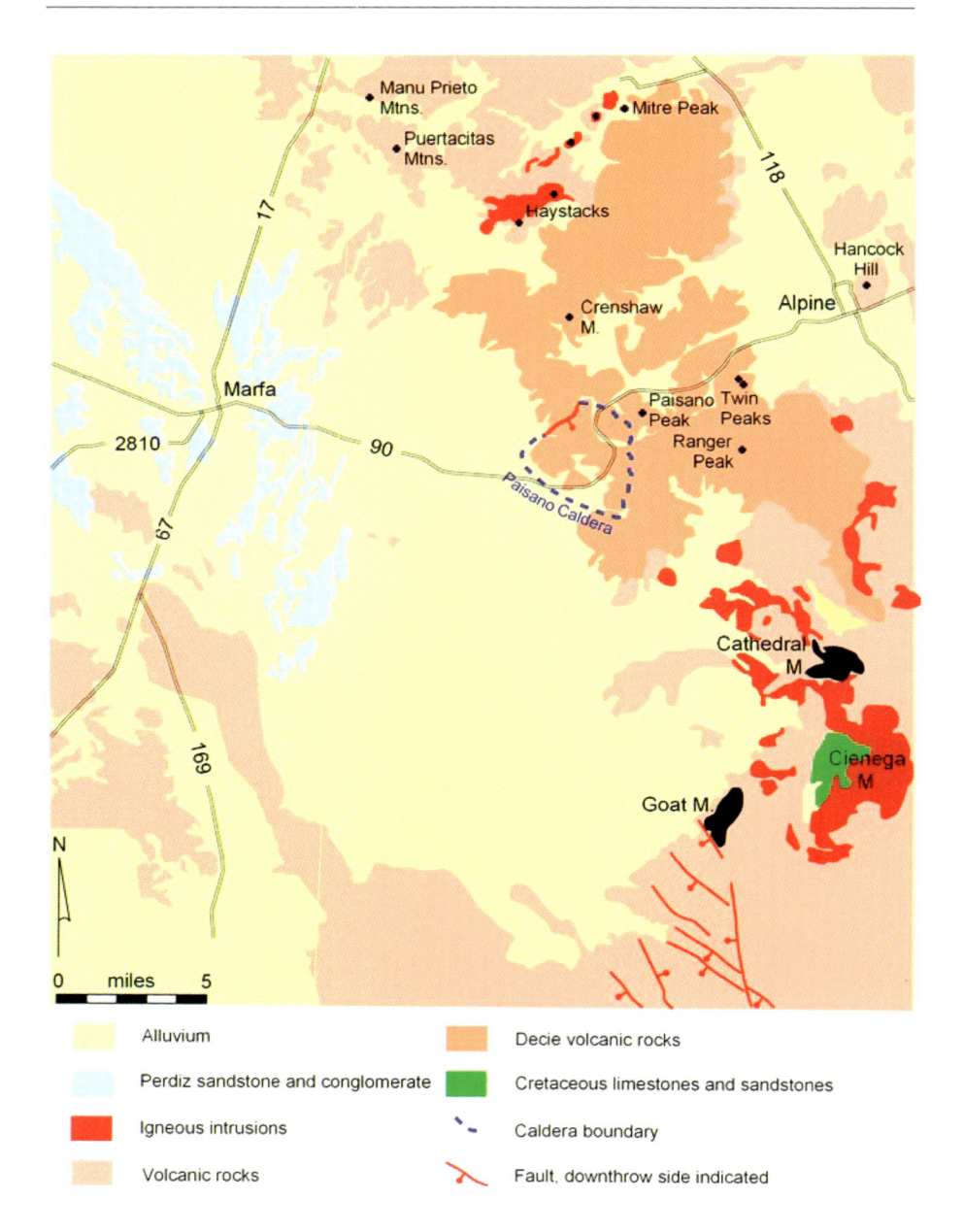

Fig 8.21: Geology of the eastern fringe of the Marfa plain, a continuation to the south of Fig. 8.1. Cathedral and Goat Mountains, although volcanic, are in black so that they can be seen on the map.

Fig. 8.22: The Presidio County courthouse was built in 1885 at a cost of $60,000, an enormous sum for those days, to a design in the Second Empire style with Italianate flourishes. It was refurbished in 2001.

Fig. 8.23: The Haystacks from Mile 3.9 north of Marfa on Highway 17 are part of a single rhyolite intrusion. The intrusion, as it pushed up from below, found two weak spots in the rocks above, thus the two mountains.

Goat Mountain to the right of Cienega Mountain at 8 o'clock is an erosional remnant similar to Cathedral Mountain, although unfaulted, and also capped by basalt. The Mitchell Mesa Welded Tuff creates a ledge about two-thirds of the way up the mountain.

The positions of Cathedral and Goat Mountains are shown in black on Fig. 8.21. They are both almost 7,000 feet high, as are the Haystacks, so it is likely that the entire area between the Haystacks and Goat Mountain was 7,000 feet above sea level at the end of the main volcanic activity.

The basalt capping Cathedral and Goat Mountains has not been dated but is younger than the Mitchell Mesa Welded Tuff (33.0 Ma) and the overlying Tascotal Formation, a volcaniclastic sedimentary apron derived from the Chinati Mountains volcanic center, and it is probably of the same age as the Sierra Vieja basalt dikes, 25-18 Ma (see page 157).

To the right of Goat Mountain, a series of faults bring the Mitchell Mesa Welded Tuff down 700 feet to become the caprock on O T Mesa about 19 miles away at 8:30 o'clock. From that point, the escarpment capped by the welded tuff slowly descends to plain level at 9:30 o'clock.

During the Second World War, 1942 to 1945, the land in front of the Viewing Area was the Marfa Army Air Field, used to train air crew, where there were as many as 30,000 men at any one time, sleeping under canvas. The runways still show up clearly on aerial photographs.

At Mile 22.8, the highway begins its descent into the valley of Alamito Creek, a valley some 6 miles wide and about 175 feet deep. This creek is the only outlet from the 1,500-square mile Marfa drainage basin. It joins the Rio Grande just south of Presidio. The Davis Mountains are very prominent on the skyline from about 1 to 2:30 o'clock, beginning with the twin white-capped Mine and White Mountains, then Brown Mountain, Paradise Mountain with the bare rock cliff of Brooks Mountain and Mount Livermore behind it, and the wedge-shaped Blue Mountain at 2:30 o'clock. These mountains are described in Chapter 7.

The intersection of US 90 and US 67 south to Presidio, and TX 17 to Fort Davis is at Mile 25.9. Turn right on to Highway 17 for downtown Marfa and Fort Davis and reset your odometer.

## Marfa

The small town of Marfa, like Alpine and many others in West Texas, began as a watering point on the Galveston, Harrisburg and San Antonio Railway in 1883. Railroad steam engines of that era had to have water every thirty miles or so. The name supposedly came from a character in a Russian novel being read at the time by the wife of one of the railroad company's executives. Marfa is the Russian for Martha.

The town quickly became a distribution point for its area and its standing was enhanced when Presidio County moved its headquarters there from Fort Davis in 1885 and the elegant courthouse was built at a cost of $60,000 in 1886 (Fig. 8.22). By 1900, the population was 900.

Marfa had a military presence for many years, beginning in 1911 when a cavalry company was stationed south of town. Activity increased during the Mexican revolution when Camp Marfa, as it was called, became headquarters for the Big Bend district and biplanes reconnoitered the border from canvas hangers south of town. This period in the area's history is beautifully documented by Smither's photographs and reminiscences in his autobiography.

Camp Marfa changed its name to Fort D. A. Russell in 1930 and became a temporary prisoner of war camp during World War II. Although the military bases were closed at the end of World War II, the Federal government again maintains a large presence through the Border Patrol which controls immigration in much of West Texas and Oklahoma from offices in Marfa.

Marfa has a growing art influence in the southwest thanks to the late Donald Judd, the sculptor and minimalist painter, who purchased the old fort in the late 1960s and transformed it into an art museum. The museum opened in 1986 and is run by the non-profit Chinati Foundation which also sponsors art and education programs.

Leaving Marfa, Highway 17 to Fort Davis turns right in front of the courthouse and then left a block ahead. The mountains to the north come into view at Mile 1.3. Blue Mountain at 11:30 o'clock, the Puertacitas Mountains (6,285 feet) at 1 o'clock (Fig. 8.24) and the Haystacks at 2 o'clock dominate the skyline (Fig. 8.23). Some of the other Davis Mountains can be seen to the left of Blue Mountain, the white-topped Mine and White Mountains at 10:30 o'clock with Brown Mountain and then Paradise Mountain to their right.

Sporadic exposures of basalt are found high on the Haystacks and Puertacitas Mountains. They were classified by Gorski as the Puertacitas Formation, a 1,250-feet thick set of basalts and tuffs that erupted from nearby vents in one of the later volcanic episodes. One vent has been identified between the Puertacitas Mountains and the Haystacks (called the Twin Mountains on USGS topographical sheets).

The two greenhouses on the left at Mile 3.8 are owned by Village Farms, a privately owned corporation based in New Jersey. It is one of the two largest growers of greenhouse tomatoes in the United States with 120 acres under glass. The company owns a third greenhouse 10 miles ahead.

Fig. 8.24: The Puertacitas Mountains in the evening sun from Mile 8.1. The mountains consist of three small intrusions set in the basalts and tuffs of the Puertacitas Formation, which originated from nearby vents. The Puertacitas Formation has not been dated but is probably late in the main volcanic cycle, perhaps as young as 30 Ma. One of the vents has been identified. The two small buttes on the left are capped by basalt, probably of Basin and Range age, 23 to 18 Ma.

Fig. 8.25: Mount Locke and Mount Fowlkes from Mile 14.6 on the skyline at dusk. The brown hills in the middle foreground are of Barrel Springs strata and lie between the Davis Mountains State Park and Blue Mountain.

The three greenhouses constitute an enormous industrial-scale enterprise, employing 500 to 600 people and producing 30,000 to 35,000 tons of tomatoes per year from a million to a million and a half plants.

The site was chosen for the availability of natural gas, used to heat the greenhouses and provide carbon dioxide and water vapor for the plants, for its winter sunshine and for the relatively cool summers. Operations shut down at the beginning of each July when the greenhouses are stripped of plants and growing medium and sterilized before replanting. The growing medium is a glass wool made from basalt and limestone which is melted and spun on disks rather like cotton candy at a fairground. Water use is about 80 million gallons per year per greenhouse, produced from the company's own wells.

A series of basalt-capped buttes crosses the highway at a saddle beginning at about Mile 9.7. The vegetation on their flanks indicates that the buttes are built up of thin flows, probably of tuff (Fig. 8.24). The basalts were most likely produced during the period of basin-and-range faulting, 23 to 18 Ma (see page 157).

About Mile 12.0, the Mano Prieto Mountains come into view on the right, a series of small hills composed of Sleeping Lion and Barrel Springs strata. Basalts crop out at road level.

At Mile 13.0, the road leaves the hills behind and descends into the drainage basin of Cienega Creek. The creek flows north to join Limpia Creek. The domes of the McDonald Observatory are now on view at about 11 o'clock, the two white domes on Mount Locke and a silver one on Mount Fowlkes (Fig.8.25).

The third Village Farms greenhouse is at Mile 16.9. The junction with Highway 166, the Davis Mountains Scenic Loop, is at Mile 18.4.

For the next two miles, a low ridge capped by Barrel Springs strata follows the highway on the right, ending in Dolores Mountain just before you enter Fort Davis.

The Jeff Davis County courthouse is at Mile 20.9.

# 9: The Salt Basin Rift

This chapter continues the geological history of the Davis Mountains area where Chapter 2 left off, at the end of the main volcanic phase around 33 Ma. Although local undated eruptions such as those in the Puertacitas Mountains continued for an unknown length of time, large-scale volcanism moved 50 miles south to the Chinati Mountains volcanic center where it continued for another 1½ million years.

The Farallon plate, which had been undergoing subduction under the North American plate since 175 Ma, reached the stage at 29 Ma where it had become almost entirely subducted. As a result its western boundary, marked by the East Pacific Rise, came up against the North American plate. Subduction halted, and the western North American plate, after being under compressive pressure for a very long time, was released from that pressure and began to extend. As it extended, faults developed to accommodate the extension, breaking the crust into fault blocks, some of which rose, some of which subsided. This created the Basin and Range topographical province (Fig. 9.1) where long mountain ranges running nearly north-south 15 to 20 miles apart are interspersed with basins 6 to 12 miles wide.

Whether the meeting of the East Pacific Rise with the North American Plate was connected with the beginning of extension is a question geologists have been asking themselves ever since plate tectonic theory was formulated in the 1970s, but it seems unlikely that it is just a coincidence.

Twenty-nine million years ago, the western United States was a wide mountain belt, 1,500 miles across and 600 miles from north to south (Fig. 9.3), rather like today's Andes. The crust had been thickened by compression in the Sevier and Laramide orogenies to as much as 40 miles in places. Western Nevada had been uplifted to 10,000 feet above sea level.

When compressive pressure eased, the thickened crust began to collapse under its own weight in places where it was warmest and hence weakest, such as over igneous intrusions in the Pacific Northwest and the northern Rocky Mountains. In other parts of the province, collapse took longer and most of the extension, in fact, did not occur until after 16 Ma.

As the thickened crust collapsed, it widened. Since 29 Ma, the western edge of the continent has moved west by 150 to 190 miles, mostly since 16 Ma. In much of the Basin and Range province, such as the sections of Southern Oregon and western Nevada comprising the Great Basin, elevations decreased.

## The Rio Grande Rift

While much of the western United States underwent Basin and Range extension, an area of the southern Rockies appears to have continued to be uplifted into what has been called the Alvarado Ridge, the high terrain on Fig. 9.3 from central New Mexico up through central Colorado. To the east of the Rockies, the landscape had been tilted, as it remains. Cretaceous strata at Killeen in central Texas, for example, are 800 feet above sea level while strata at the same horizon near El Paso are 4,700 feet above sea level.

At about the start of Basin and Range extension, the Rio Grande Rift began developing up the crest of the Alvarado Ridge. A rift, in geological parlance, is a long, narrow trough where the entire lithosphere has been deformed while being extended or stretched. Sometimes a rift is a precursor to a continent splitting in two. The Red Sea, for example, began as a rift that now separates Africa from Saudi Arabia. Others remain as rift valleys, the Rhine valley in Europe or the several rift valleys of East Africa, for example.

Both the uplift and the rift are thought to have been produced by upwelling currents in the asthenosphere, perhaps connected to the subduction of the Farallon plate. Such currents send heat and mass up, creating magma chambers in the upper asthenosphere. This causes the lithosphere to uplift and form broad arches in which the crust is thinned. Faulting then occurs along the uplifted crest in to which magma oozes, leading to further uplift as the temperature increases. Finally, you have deep rifts composed of a series of grabens whose bottoms are moving down compared to the arches but moving up relative to the convecting magmas below.

A recently-completed seismic geophysical investigation called La Ristra provided evidence that the Rio Grande Rift was caused by this process.

150                                    9: THE SALT BASIN RIFT

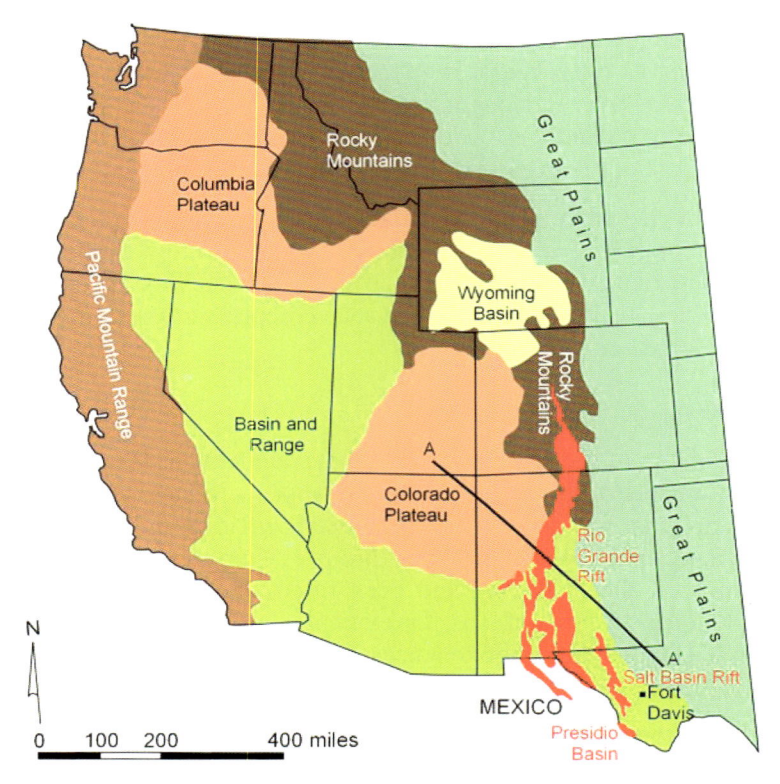

Fig. 9.1: Physiographic provinces of the western United States with the Rio Grande and Salt Basin Rifts superimposed. The La Ristra line is along A-A'.

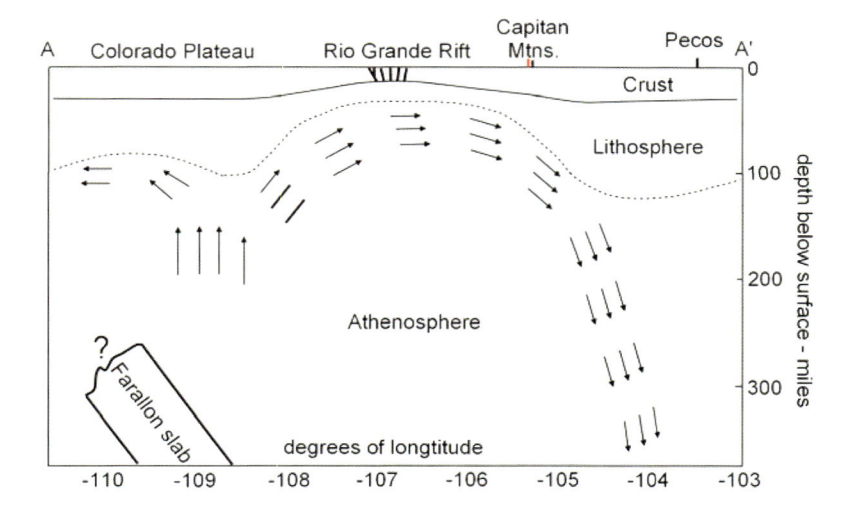

Fig. 9.2: Cross-section along the La Ristra line, modified from Goa et al.

# 9: THE SALT BASIN RIFT

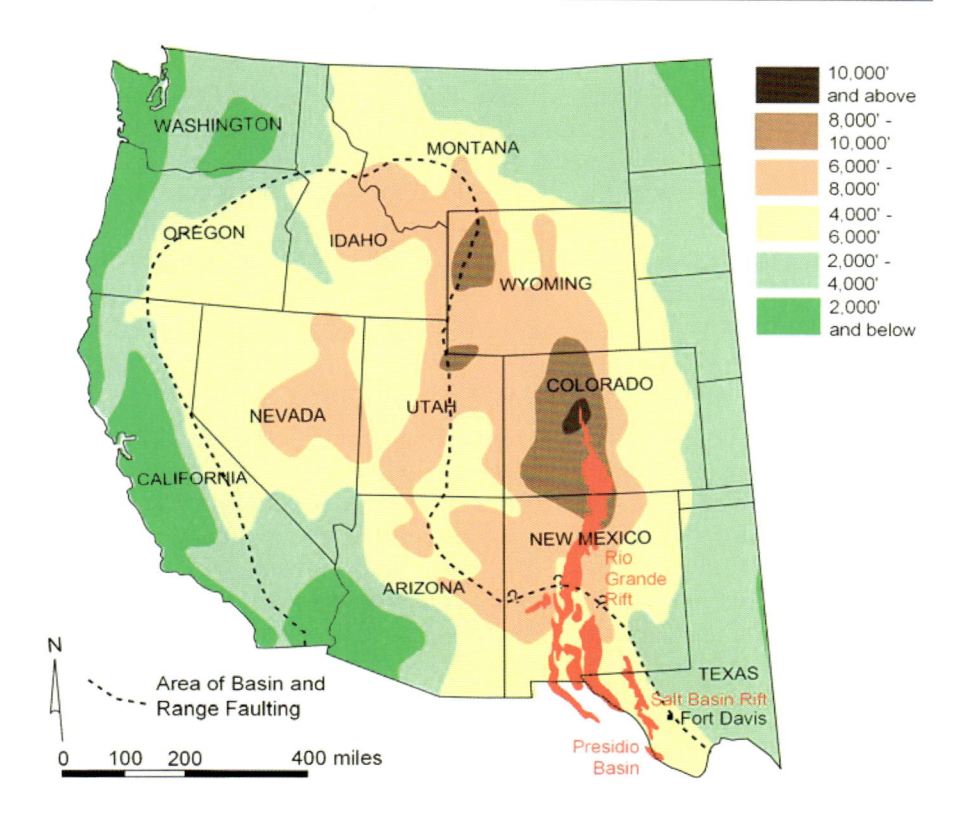

Fig. 9.3: The Rio Grande and Salt Basin Rifts superimposed on a map of Western U.S. elevations. Note how the Rio Grande Rift coincides with the crest of the southern Rocky Mountains ridge, the Alvarado Ridge.

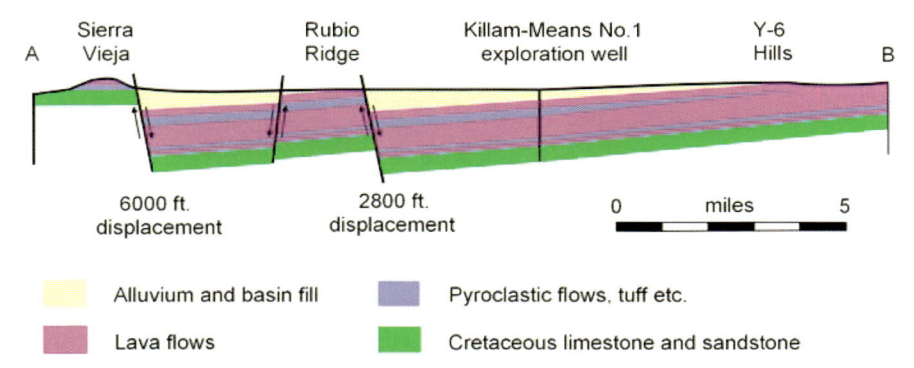

Fig. 9.4: This section across the Salt Basin Rift from the Y-6 Hills to the Sierra Vieja shows strata sloping into a half-graben, faulted at the foothills of the Sierra Vieja, with Rubio Ridge as a horst in the graben. The line of section is shown on Fig. 10.1.

In the investigation, a series of seismic sensors were laid across the rift in 1999 from Pecos in Texas to Lake Powell in Utah (line A-A' on Fig. 9.1) and left for about 18 months. The instruments recorded shock waves from earthquakes of greater than 5.6 magnitude on the Richter scale that occurred anywhere on Earth, 29 in all. Just as ultrasound machines can construct images of internal organs by analyzing the speed at which sound waves pass through a body, so can computers construct images of the Earth's internal structure by analyzing the speed at which shock waves from earthquakes pass through the earth, in a process called tomographic reconstruction. Waves pass through dense material quicker than less dense material and through cold material quicker than hot material.

Papers based on the La Ristra research are being published as this book is being written. Fig. 9.2 is a simplified version of a tomographic reconstruction along the La Ristra line from a paper by Gao, Grande, Baldridge, Wilson, West, Ni, and Aster. It shows mantle currents upwelling from an anomaly in the lower left, flowing west across the Colorado Plateau and east across the rift and down near Pecos at the edge of the Great Plains. The authors suggest that the lower left anomaly is a remnant of the Farallon Plate trailing edge, shown in the figure as a cartoon image (this data point was used in constructing Fig. 2.4).

The diagram also shows that the crust is 22 miles thick under the rift, thinner than under the Colorado Plateau or the Great Plains, where it is 27 and 28 miles thick, respectively. Other studies have shown that the crust is less than 17 miles thick under the rift near Truth or Consequences, N.M., 100 miles south of the La Ristra line. The lithosphere, including the crust, has been thinned under the rift to about 30 miles compared to 124 miles under the Great Plains and 75 to 90 miles under the Colorado Plateau.

The Rio Grande Rift has been intensely studied over the past 30 years and its history well established. It developed quite slowly; by 25 Ma, shallow basins had developed along its length in which volcanic ash accumulated. For the next 10 million years it remained broad and shallow until 15 Ma when uplifting and block faulting began to produce the rift as it appears today; uplift averaged between 5 and 10 inches per thousand years. It now consists of 4 major linked basins up to 26,000 feet deep, filled with lake and river sediments and is more or less dormant.

Extension across the rift, i.e. the amount by which the Earth's crust was stretched, varies from about 10 per cent in Colorado, 28 per cent at Albuquerque and as much as 50 per cent near the Mexican border where the rift branches into multiple basins and uplifted blocks. The southern boundary of the rift is uncertain. It turns abruptly to the southeast at El Paso along the axis of the Jurassic-Cretaceous Chihuahua Trough and may continue as far as the deep basin at Presidio (Fig 9.3).

## The Salt Basin Rift

A series of basins parallel to the Rio Grande Rift occurs from the Guadalupe Mountains in New Mexico to Marfa with associated faulting continuing up to the Capitan Mountains in New Mexico. In this book, for convenience, this feature is called the Salt Basin Rift, after one of the larger basins, although whether it meets the criteria to be called a rift is in question. The La Ristra line passed just north of the Capitan Mountains (Fig. 9.2) and the mountains appear to be on the eastern flank of the bulge in the asthenosphere, so it seems probable that the Salt Basin Rift has been created by the same processes as the Rio Grande Rift. It also appears to be of roughly the same age, i.e. it developed mainly after 15 Ma.

The Salt Basin Rift is a series of connected grabens or half-grabens. The grabens have been filled with lake and stream deposits of silt, gravel and sand eroded from the adjacent mountains, up to 2,000 feet thick in the Salt Basin north of Van Horn and 1,200 feet thick around Valentine. Volcanism that occurred in the last 5 million years in the Rio Grande Rift area did not occur, however, causing Seager and Morgan to propose that this is a rift in early stages of development.

The Salt Basin Rift has had a profound effect on the Davis Mountains. Geologists mapping in the mountains, Anderson among others, have reported that strata in the mountains dip to the southwest. Using surface outcrops and data from oil exploration wells, the base of the volcanic rocks has been plotted on Fig. 9.5. Data are sparse and there is probably faulting in the basin fill not shown on the map, but it is clear that the Davis Mountains and the strata underlying them to the southwest of the Star Mountain anticline have been tilted into the rift.

The deepest basin along the Davis Mountains is in the vicinity of Valentine, where the Killam-Means oil exploration well, drilled about a mile north of Valentine, intersected 1,100 feet of alluvium and 5,400 feet of volcanic rocks before reaching the top of the Cretaceous strata at 2,097 feet below sea level (Fig. 9.7), some 4,000 feet below their highest level in the Davis Mountains.

## Basalt Volcanic Activity

Basalt lavas erupted widely throughout the Basin and Range province as faulting provided conduits for magmas to reach the surface. Henry, Kunk and McIntosh proposed that basaltic magma was present throughout the entire volcanic period in the Davis Mountains but was overlain by less dense felsic magma and could not rise to the surface except when the latter cooled and solidified. Presumably, by the time of Basin and Range faulting, the overlying felsic magma had been all erupted.

154    9: THE SALT BASIN RIFT

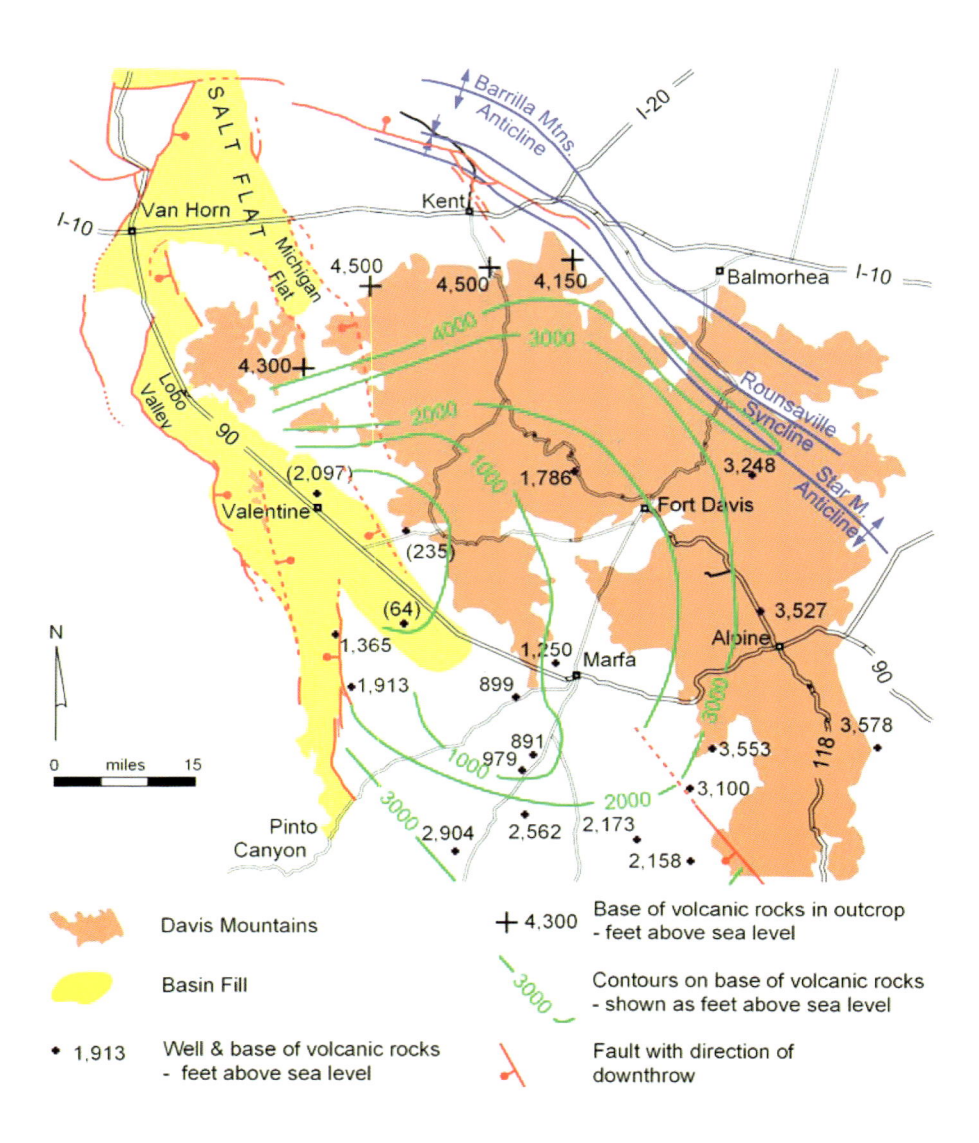

Fig. 9.5: From the Star Mountain anticline, the base of the volcanic rocks slopes down towards the Salt Basin, a set of basins along a rift valley that ends somewhere in the vicinity of Marfa.

# 9: THE SALT BASIN RIFT

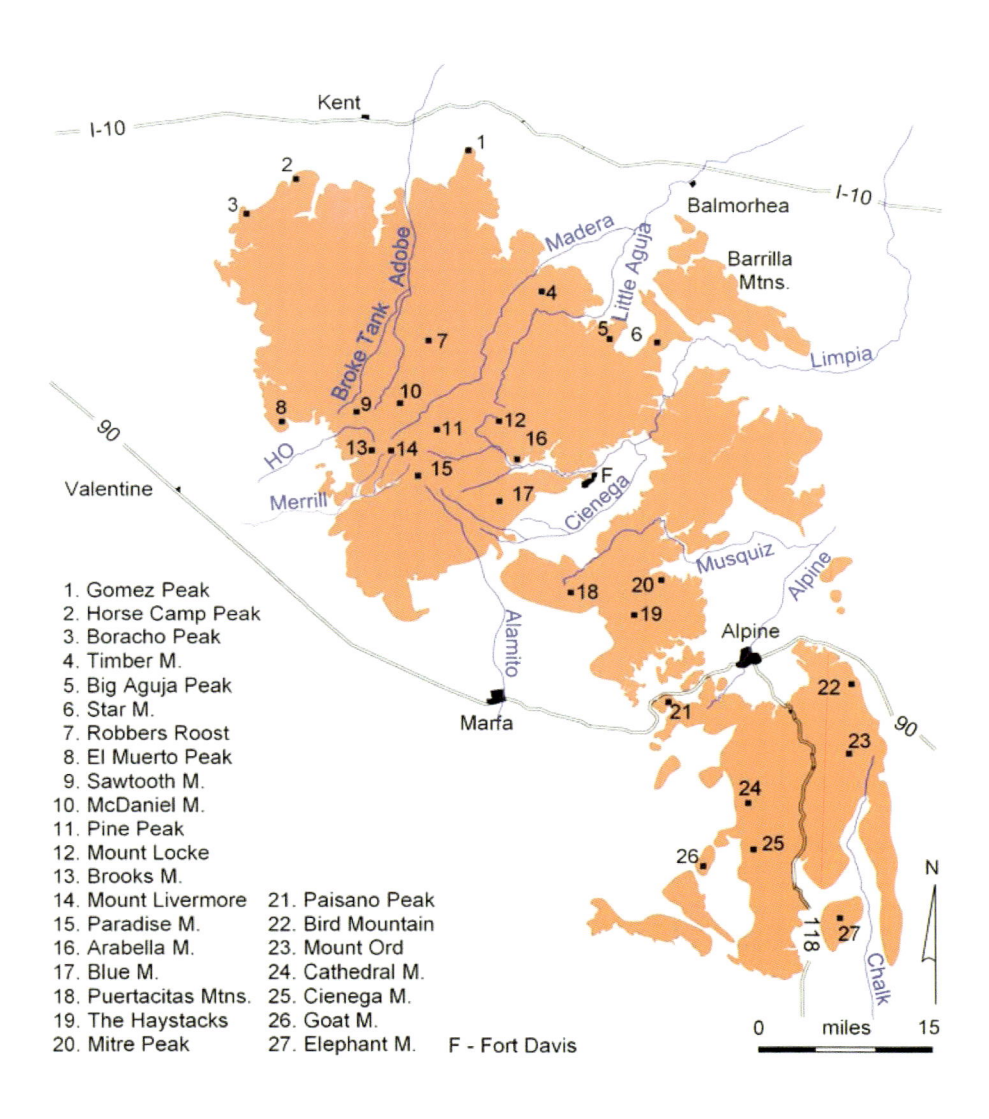

1. Gomez Peak
2. Horse Camp Peak
3. Boracho Peak
4. Timber M.
5. Big Aguja Peak
6. Star M.
7. Robbers Roost
8. El Muerto Peak
9. Sawtooth M.
10. McDaniel M.
11. Pine Peak
12. Mount Locke
13. Brooks M.
14. Mount Livermore
15. Paradise M.
16. Arabella M.
17. Blue M.
18. Puertacitas Mtns.
19. The Haystacks
20. Mitre Peak
21. Paisano Peak
22. Bird Mountain
23. Mount Ord
24. Cathedral M.
25. Cienega M.
26. Goat M.
27. Elephant M.

F - Fort Davis

Fig. 9.6: Drainage patterns in the Davis Mountains radiate from the Mount Livermore-Brooks Mountain area. Note how close the Alamito Creek origin is to the Madera and Limpia Creek origin, although the former drains south to the Rio Grande and the latter to the Pecos River.

| Type of Rock | Thickness - feet | Tentative Correlation |
|---|---|---|
| Surface alluvium and basin fill – samples missing or questionable | 1,130 | |
| Basalt, trachybasalt, trachyte | 560 | |
| Basalt | 540 | |
| Tuff | 790 | |
| Basalt | 30 | |
| Rhyolite, several flows separated by 3 basal tuffs and a vitrophyre | 1,530 | |
| Basalt, 3 or 4 flows, according to electric logs | 390 | |
| Rhyolite, at least 3 flows | 350 | |
| Rhyolite, 3 flows | 270 | |
| Tuff | 310 | |
| Basalt | 440 | |
| Tuff with some sandstone and shale beds | 220 | Huelster Formation |
| Limestone | 1,040 | Boquillas - Boracho Limestones |
| Sandstone with thin beds of sandstone & shale | 770 | Cox Sandstone |
| Total | 8,370 | |

Fig. 9.7: Strata in the Killam Means No.1 well (adapted from Woodward)

An interesting insight into the later volcanic activity is given by the Killam-Means well. In it, basin fill protected the volcanic rocks below from erosion and hence they are better preserved than those remaining on the surface these many million years. Cores from the well were studied for a master's thesis by J.E. Woodward at the University of Texas in 1953. His descriptions, summarized in Fig. 9.7, are from well cuttings. In well cuttings, small rock chip samples are taken every 10 feet, and give you less information than you would get from surface exposures or drill core, but some generalizations can be made.

The lowest basalt, 440 feet thick, is thicker than most basalts at this level but otherwise the sequence up to the upper two basalts is similar to that in the Davis Mountains. The upper basalts were correlated by Wightman to basalts mapped on the Y6 Hills about 7½ miles north of the well and to similar rocks cropping out in the seven-mile long Rubio Ridge west of the highway where he identified a conical hill as a possible vent for the basalts. The 790-feet tuff below the upper basalts in the well is missing in the Y-6 Hills. This suggests that by the time the tuffs were deposited, the Valentine basin had sunk enough to protect them from erosion until the overlying basalts erupted. In the Y-6 Hills area, which is not in the basin, the tuff was eroded away during this interval.

# 9: THE SALT BASIN RIFT

The structure across the basin from the Y-6 Hills through the Killam well and Rubio Ridge to the Sierra Vieja is shown in Fig. 9.4. Water wells west of the ridge were drilled through 500 to 600 feet of alluvium and basin fill before meeting bedrock, leading me to conclude that the ridge is a horst. I have also assumed that the basin is a half-graben, faulted on its west or Sierra Vieja side, with strata dipping down to the west at about 6 degrees, the dip that Wightman measured in the Y-6 Hills.

None of the younger basalts around the Davis Mountains have ever been dated but in Big Bend Ranch State Park, 60 miles to the south, Christopher Henry describes dated basalts that erupted during the early part of basin-and-range faulting, 25 to 18 Ma. Similarly, a swarm of basaltic dikes in the Cretaceous strata on the western side of Sierra Vieja were dated at between 18 and 23 million years by Dasch, Armstrong and Clabaugh. There, erosion has probably carried off the overlying basalt flows. It seems likely that the basalts in and around the Salt Basin Rift are about the same age as in those two locations.

## Current Landscapes

Since the volcanic eruptions ended around 30 million years ago, the mountains have been eroded into their current configuration (Fig. 9.6). In the main mountain section, canyons radiate from the Mount Livermore-Brooks Mountain dome (Fig. 2.14). The plateau to its south, which extends from Paradise Mountain (7,719 feet) to Pine Peak (7,710 feet), is being eroded by Limpia Creek and its tributaries, attacking from the northeast. One of the Limpia Creek tributaries, Jones Creek, now extends to the foothills of Blue Mountain, which it isolates from the plateau. Limpia Creek itself rises in the foothills of Mount Livermore.

The section northeast of Highway 118 is also a plateau, about 6,500 feet high, in which canyons, the deepest of which is Madera Canyon, are cutting back into the mountains from the north. Madera Canyon now extends to the foothills of Mount Livermore.

South of Blue Mountain, the volcanic rocks form a plateau at about 5,500 feet above sea level, 1,000 feet lower than the northeast sector, probably because the younger volcanic units are absent. However, that is not to say that they were not present at one time. Intrusions in the Alpine area, which were entirely underground when intruded at perhaps 33 or 34 million years ago, rise up to almost 7,000 feet, so it is probable that the volcanic landscape was at least that high when they were intruded. Cathedral and Goat Mountains provide further evidence that the area between the Chinati and Paisano calderas had been filled up to the same level with debris eroded off the Chinati Mountains by the end of volcanic activity (see page 144).

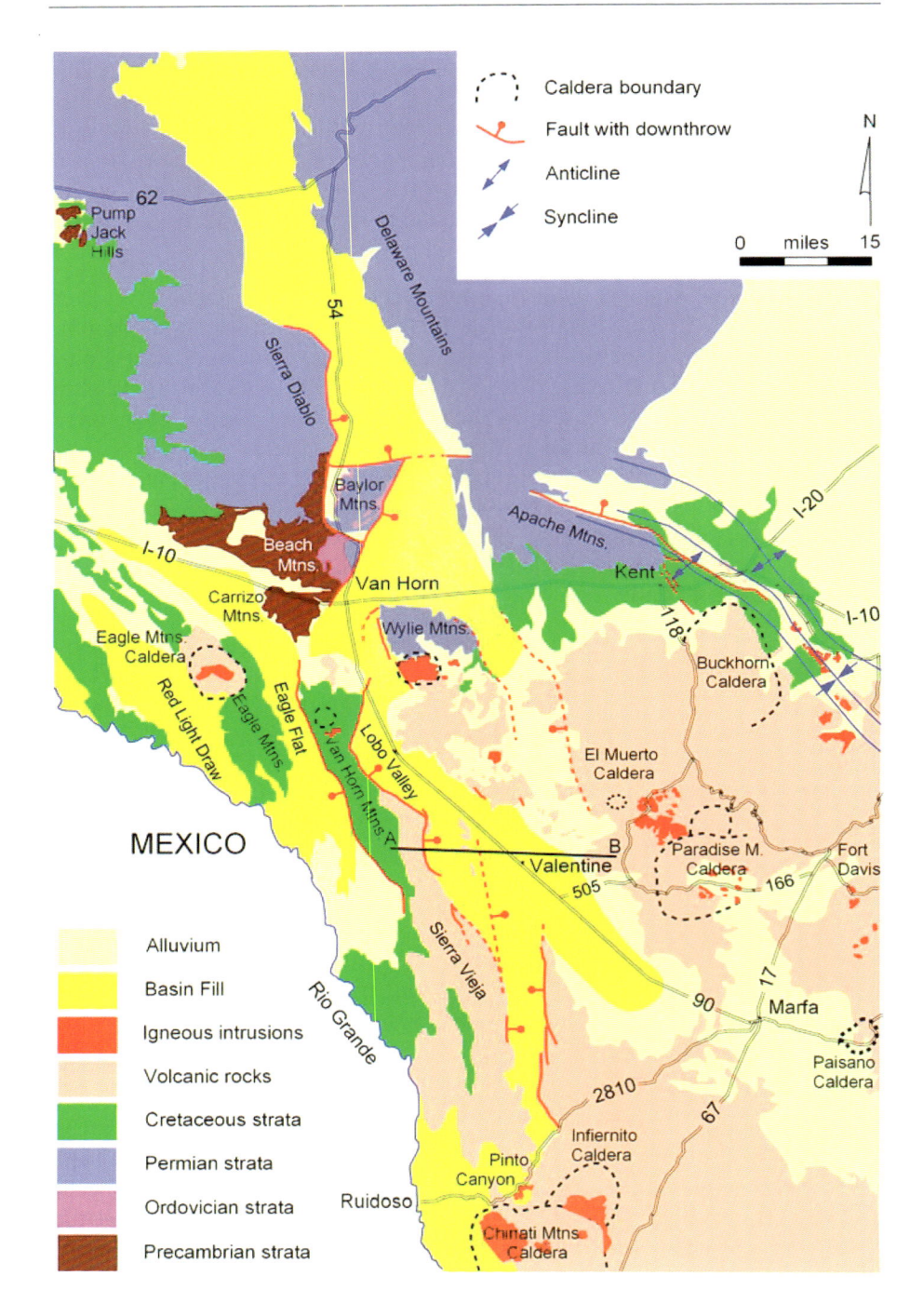

Fig. 10.1: Geology of the western Davis Mountains and the Salt Basin Rift

# 10: Kent – Van Horn – Valentine – Marfa

## 109 miles

The junction of TX 118 and Interstate 10 is at Mile 176 on the interstate. Follow the signs for I-10 West. Stops along the interstate are at exits. Mileages to Van Horn are measured from the junction.

Kent, like Van Horn began life as a watering stop on the railroad built by the Texas and Pacific Railway Company, which was given a federal charter in 1871 to build from Marshall, Texas, to San Diego, California. The company was taken over by the Missouri Pacific in 1976 and Union Pacific in 1982.

The interstate runs on Cretaceous strata, mainly at the Boracho Limestone level (the full succession of strata in this area is given in Fig. 10.2), while towards the Apache Mountains on the right progressively older rocks, from the Finlay through the Seven Rivers Formations, are exposed in a large asymmetrical anticline. The crest of the anticline forms the long straight ridge of the Apache Mountains on the skyline to the right.

Capitan Limestone, a Permian fossil barrier reef, crops out at the crest of the anticline. The reef was a submerged ridge that formed over as much as 10 million years by growth and accumulation of the invertebrate skeletons of algae, sponges and tiny colonial animals called bryozoans, now extinct. The skeletons were encrusted by organisms that grew over and cemented the solid reef rock, unlike the rigid coral framework of modern reefs.

The formation takes its name from El Capitan in the southern Guadalupe Mountains, where it forms a cliff 1,000 feet high that can be seen from Interstate 10 ahead. The reef crops out for over 50 miles along the Guadalupe Mountains (Fig. 10.3), in the Apache Mountains and the Glass Mountains northeast of Alpine. It has been tracked underground between the Guadalupe and the Glass Mountains by oil exploration well data. Geologists believe that one or both of the two gaps in the reef, north of Alpine and between the Apache and Guadalupe Mountains was an inlet from the open ocean to the Delaware Basin.

| Period | Age m.y. | Formation | Description |
|---|---|---|---|
| Upper Cretaceous | 98-45 | Undivided | Marl, shale and clayey limestone; 300 ft. thick. |
| Lower Cretaceous | 144-98 | Boquillas | Upper part, interbedded marl and shale; lower part, limestone, silty to sandy, flaggy, dark grayish orange near base; marine fossils; 200 ft. thick. |
| | | Buda Limestone | Limestone, thin to thick bedded, middle 60 ft. clayey; sandstone locally at base; 140 ft. thick. |
| | | Boracho | Two members; *San Martine Limestone:* clayey limestone, thin to thick bedded with calcareous shale interbeds; marine fossils; 230 ft. thick. |
| | | | *Levinson Limestone:* Upper two-thirds limestone, light to light olive gray marl with interbeds of yellowish marl and shale; lower third shale, light to dark gray with occasional limestone beds; abundant marine fossils; 150 ft. thick. |
| | | Finlay | Sandstone, sandy limestone and massive coarse-grained limestone, 15-40 ft. thick. |
| | | Cox Sandstone | Sandstone, fine to medium grained, interbeds of silty marl and quartz-pebble conglomerate; 25-170 ft. thick. |
| | | Yearwood | Limestone with interbedded shale, conglomerate at base up to 55 ft. thick; 160 ft. thick, thins to north. |
| Permian | 286-245 | Capitan Limestone | Reef limestone and dolomite, medium to coarse grained, massive, beds 15-100 ft. thick, white, light gray, grayish yellow, brownish yellow; marine fossils; up to 900 ft. thick. |
| | | Tansill | Dolomite, fine grained, very thick to medium bedded, pale yellowish brown; 75-100 ft. thick. |
| | | Yates | Siltstone, shale, limestone and dolomite; up to 290 ft. thick. |
| | | Seven Rivers | Dolomite, fine-grained, light gray to very pale orange; 440 ft. thick. |
| | | Victorio Peak Limestone | Limestone, thick bedded, light gray along east escarpment of Sierra Diablo, 900-1,500 ft. thick. In Wylie Mountains – dolomite; marl, poorly exposed, creates slope break; limestone, marl interbeds near base; marly limestone; 1,600 ft. thick. |
| | | Bone Spring | Mostly limestone, thin bedded, dark gray to black, some beds of shale and shaly limestone; 900-1,700 ft. thick. |
| | | Hueco Limestone | In Sierra Diablo - dolomitic limestone and limestone, dark gray, cliff-forming; shale, conglomerate and sandstone. In Wylie and Eagle Mountains – limestone gray, cliff forming; grades down into shale, siltstone, sandstone and conglomerate; 200 ft. thick. |

Fig. 10.2: Sedimentary strata along the northern Davis Mountains and the Apache Mountains.

A sedimentary ridge follows the interstate on the left, capped by a Buda limestone ledge. The volcanic hills beyond it, chiefly Horse Camp Peak (5,892 feet) about 5 miles away, can be seen through a gap in the ridge formed by Hurds Draw at Mile 172. Boracho Limestone is exposed in road cuts on both sides of the highway for 12 miles from Mile 2.2. The overlying Buda Formation forms the caprock on the low hills just south of the highway but does not appear in road cuts.

At Mile 7.0 on the right, cuestas of Yates and Seven Rivers strata dip off the anticline to the southwest. From the overpass at the Boracho exit (Mile 9.4) the Sierra Diablo Mountains are on the skyline from about 12 to 2 o'clock rising up to 3,000 feet above the flat below, with great thick limestone beds of the Permian Victorio Peak, Hueco and Bone Spring Formations cropping out on their flanks. You can also see Boracho Peak (5,652 feet) at 9 o'clock from the exit, another of the Davis Mountains volcanic hills.

About Mile 11.0, the highway rounds a corner among road cuts of well-bedded Boracho Limestone and the mountains of the Van Horn area come into view. The Eagle Mountains on the far horizon at 12 o'clock are about 16 miles southwest of Van Horn. The highest point in the range is Eagle Peak (7,484 feet). The brown Wylie Mountains are to their front south of the interstate.

Just before the Plateau exit, the craggy hills at 9 o'clock are outcrops of Permian Seven Rivers dolomite, a magnesium limestone, on the east side of the Michigan Flat graben. Black Peak, in front, is a Cox Sandstone outlier standing up above the flat, protected from erosion by a small intrusion that forms its rugged cap.

## Stop 1: Plateau Exit – The Salt Basin Rift (Mile 16.0)

The overpass at this exit is a convenient point to view the flats of the Salt Basin Rift which stretch to the north as far as the Guadalupe Mountains and to the south down Michigan Flat at 9 o'clock. The rift is made up of a series of grabens, or basins, along a trend from the Guadalupe Mountains in New Mexico to the vicinity of Marfa, in which strata have dropped down relative to their surroundings (see page 153). It has partially filled with a distinctive alluvium called basin fill, composed of lake and river deposits of clay, silt and sand such as the cross-bedded sandstones and conglomerates exposed in the road cut under the overpass. The rift splits into two around the Wylie Mountains. The boundary of the eastern wing, Michigan Flat, is about a mile back, hidden in alluvium.

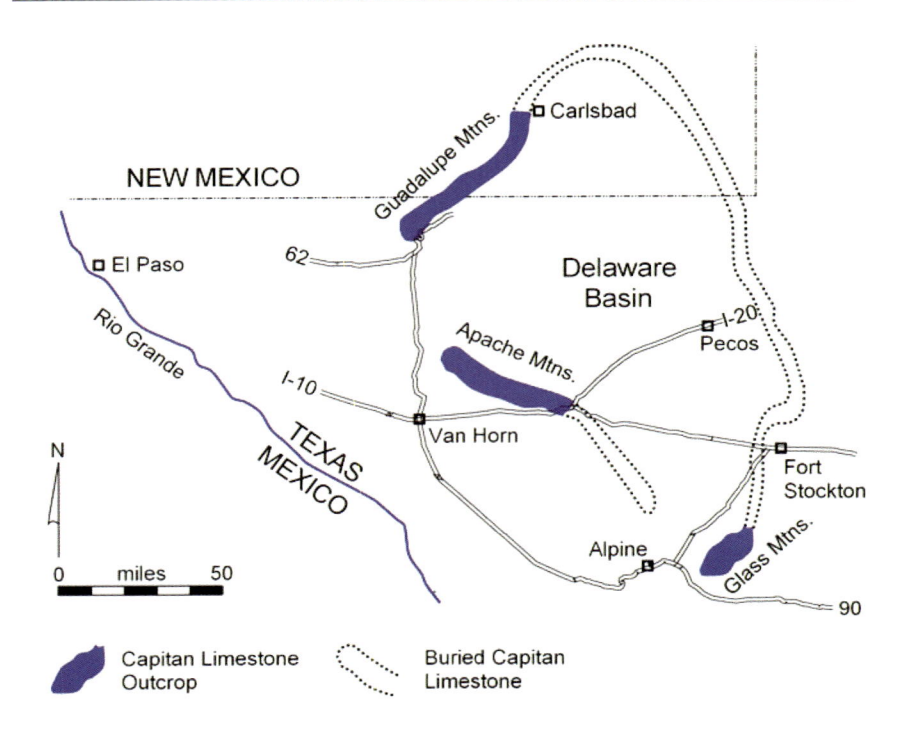

Fig. 10.3: The Capitan Reef as it appears today. The gap between the Guadalupe and Apache Mountains is perhaps the location of the inlet from the open ocean to the Delaware Basin. Another candidate is the gap between the Glass and the Apache Mountains (see also Fig. 2.6).

Gravity anomaly maps help explain the rift's structure. Gravity anomalies compare gravity values as measured by a gravimeter to theoretical values calculated for a standard model of the Earth. The differences are plotted as point values and then linked by lines of equal value. They can be negative or positive. Negative values point out areas where the Earth's subsurface has lower gravity than the standard model, positive values the opposite. Gravity anomalies show where lower density rocks such as limestones and alluvium are thicker than normal and highlight the structures of rifts.

A gravimeter is quite small and portable, about the size of a large thermos flask. It uses a mass suspended from a sensitive spring and a very accurate measuring system to measure the extension of the spring as the acceleration due to gravity increases or decreases. The unit of measurement is the gal (named for Galileo) which is equal to one centimeter per second per second, usually plotted on maps as milligals, one-thousands of a gal.

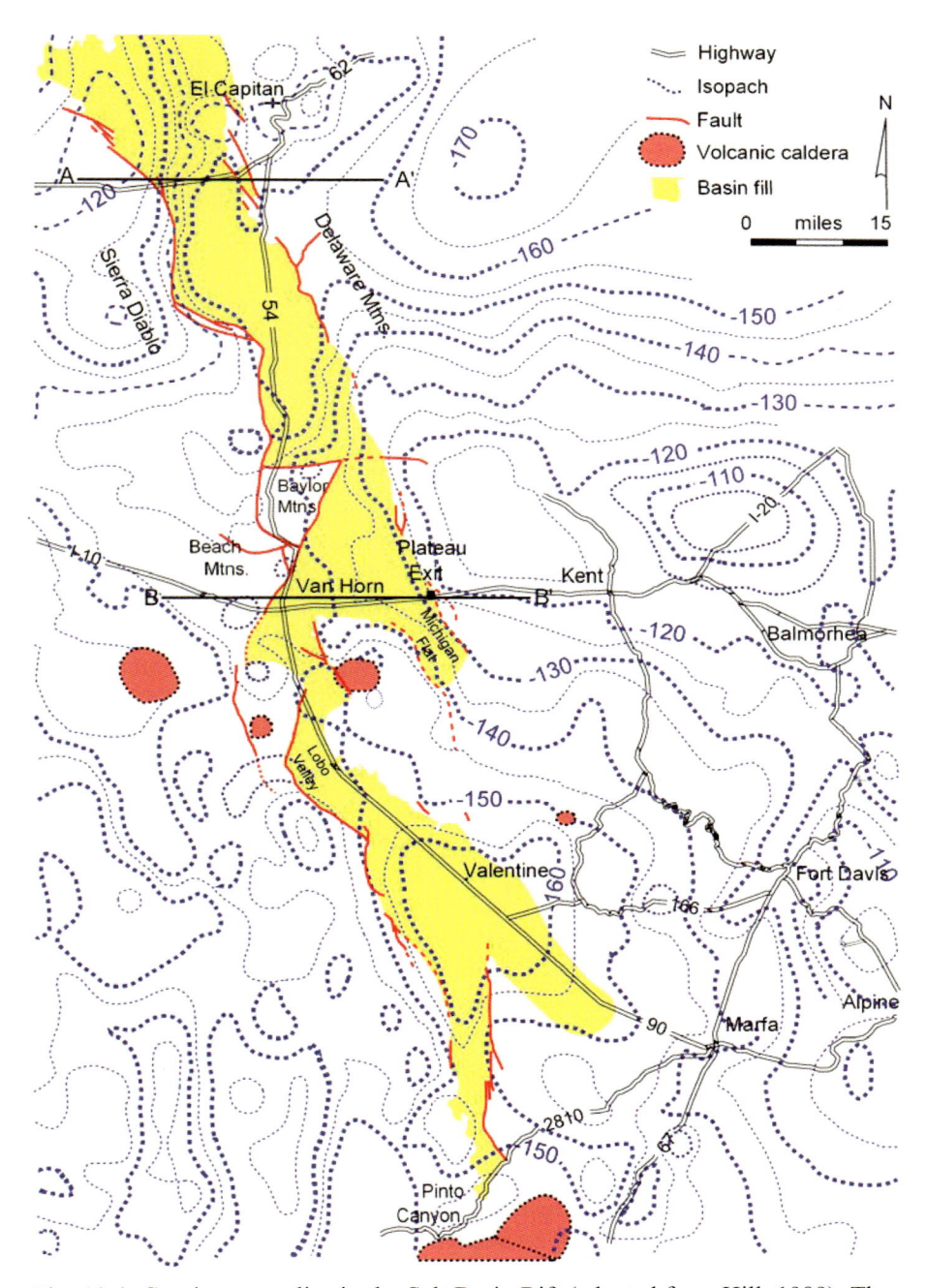

Fig. 10.4: Gravity anomalies in the Salt Basin Rift (adapted from Hill, 1999). The largest anomalies are in the salt Basin north of Van Horn and around Valentine. The -150 mgal contour in the latter suggests that the southern branch of the rift continues across to Highway 67 south of Marfa while the other branch dies out northwest of Marfa.

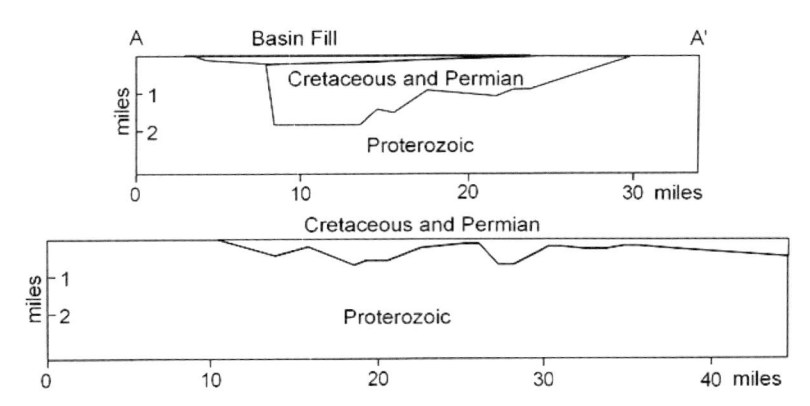

Fig. 10.5: Sections across the northern Salt Basin Rift (adapted from Keller and Peeples). The lines of section are shown in Fig. 10.4.

Maps of anomalies in the Earth's gravitational field have been constructed along the Salt Basin Rift and, indeed, for of all West Texas, by Randy Keller and his colleagues at the University of Texas at El Paso, and are being continually revised. Fig. 10.4 shows anomalies along the rift. The rift takes its name from the largest basin, the Salt Basin which stretches for about 30 miles south from Highway 62 as outlined by the -150 milligal contour.

The Salt Basin name comes from the salt that is found there. Before the advent of modern subsurface salt mining in Texas by the Morton Salt Company in 1929, salt for domestic use was mainly derived from salt flats such as these. Salt builds up because the rift north of Interstate 10 has no external outlet and rainfall accumulates in temporary or ephemeral lakes. When the water dries up it leaves *playas*, flat, vegetation-free areas that are found in areas of low rainfall and high evaporation rates. Water from the surrounding areas, that has percolated through Permian sediments containing a certain amount of evaporites, mostly salt and gypsum, is drawn up to the playa surface where the salt is concentrated by evaporation.

Other features worth mentioning on the gravity map are the large low in the Delaware Basin and moderate highs in the Sierra Diablo and the Apache Mountains, where dense Proterozoic rocks are near the surface. The intense high east of the Apache Mountains is in the Delaware Basin and is thought to be due to a shallow igneous intrusion.

The cross-sections in Fig. 10.5 were also calculated from gravity data by Keller and Peeples. The northern one shows that along Highway 62, the base of the Permian rocks on the west of the basin is nearly 10,000 feet lower than on the east, with the basin floor being tilted to the west. Geologists call this type of structure a half-graben, one that is faulted on only one side, in this case the west side. The second section along Interstate 10, B-B', shows that the basin is shallower here, about 3,600 feet at most, and almost disappears opposite the Wylie Mountains.

Continuing along Interstate 10, the brownish-colored Baylor Mountains are at 1 o'clock from the Michigan Flat exit at Mile 23.0. They, like the Wylie Mountains ahead, are composed of Permian Bone Spring and Hueco limestones that support little vegetation over a surface of limestone rubble. The rubble turns grey-brown as it weathers, giving the mountains their color. The high Davis Mountains around Mount Livermore are on the skyline at about 8 o'clock.

At Mile 25.0 on the left, the low eastern foothills of the Wylie Mountains are dimpled, a characteristic feature of lower Permian limestone terrain. Further along, as the hills get higher, limestone, dolomite and marl beds of the Permian Victorio Formation crop out intermittently. They dip to the east, as does the entire Wylie Mountains block (Fig. 10.7).

## Stop 2: Wild Horse Exit (Mile 30.0)

The overpass at this exit makes an excellent viewpoint across the Salt Basin Rift to the Van Horn area ahead. The mountains around the town are on an uplift of a section of the Earth's crust, called the Diablo Platform by petroleum geologists. Fig. 10.6 is a contoured map of the upper surface of Proterozoic rocks in the area. This surface forms a ridge rising 10,000 feet in the 75 miles from Marfa to the Carrizo Mountains at Van Horn and then levels out to between 5,000 and 7,500 feet above its level at Marfa. Proterozoic rocks also crop out in the Franklin Mountains at El Paso in a raised block in the Rio Grande Rift that is probably separate from the Diablo uplift.

Proterozoic rocks drop steeply on either side of the uplift, into the Delaware Basin on the northeast where they are as much as 30,000 feet under sea level, and into the Chihuahua Trough on the southwest where they are believed to be as much as 40,000 feet below sea level.

This enormous feature has been rising on average at about one inch per thousand years since at least the Permian period, 250 million years ago, and is still rising. According to Reilinger, Brown and Oliver, uplift of about 5 inches occurred along the El Paso-Carlsbad road between 1934 and 1958, 200 times faster than the inch per thousand year average.

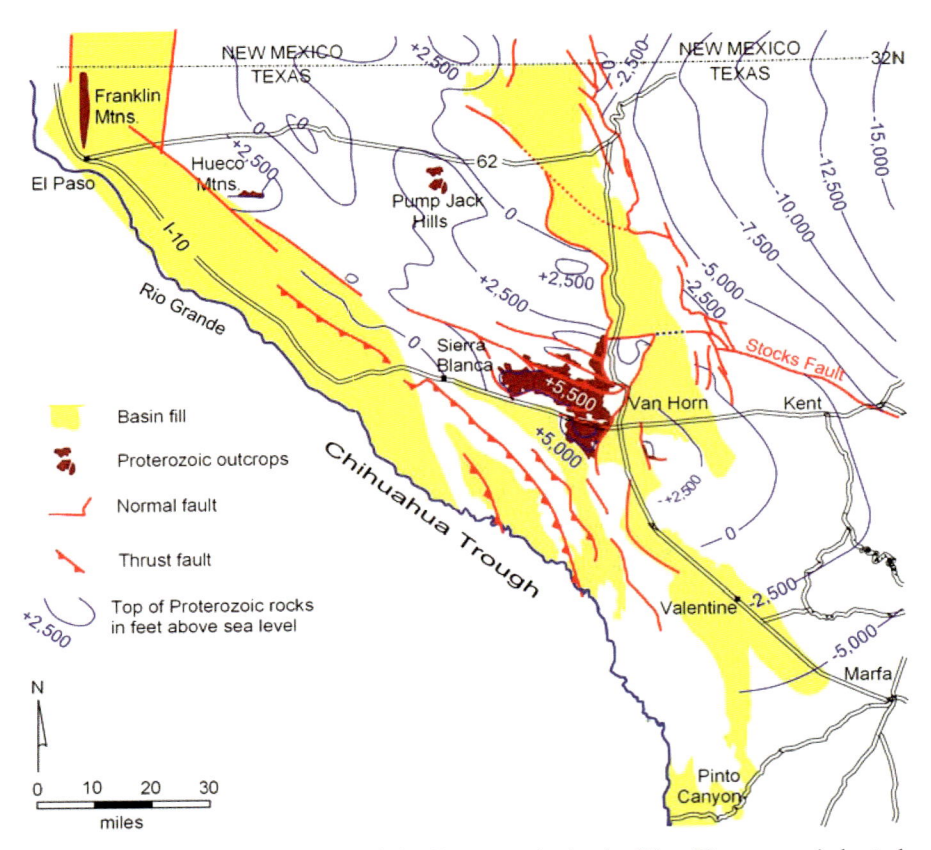

Fig. 10.6: Contours of the top of the Proterozoic in the Van Horn area (adapted from King and Flawn). An uplift, called by petroleum geologists the Diablo Platform, runs from top left to bottom right across the map. Proterozoic rocks outcrop in several places along the uplift, including the Carrizo Mountains just west of Van Horn. They are also exposed in the Franklin Mountains, a horst in the Rio Grande Rift north of El Paso.

Why did it rise? No one knows for certain. Uplifts are caused by upwelling currents in the asthenosphere and geophysical research has shown that such currents exist across the Rio Grande Rift (page 149), not that far away, although the Diablo uplift appears to have begun much earlier. In any event, it has had an enormous impact on the scenery around Van Horn where most of the mountains on view, the Baylor Mountains (5,564 feet) at 2 o'clock, the Sierra Diablo (6,628 feet) at 1:30 o'clock, the Beach Mountains (5,827 feet) at 1 o'clock, the Wylie Mountains (5,264 feet) at 9 o'clock, are uplifted blocks above the town (4,047 feet).

The rest area at Mile 31.6 is another good point to view the Van Horn scenery.

Fig. 10.7: The Wylie Mountains at daybreak from Mile 29.5. The mountains have the characteristic dimpled look of the older Permian limestones, also seen in the northern Van Horn Mountains. Outcrops are uncommon.

Fig. 10.8: The peak at the southern end of the Beach Mountains at daybreak from Van Horn. The mountains are a 5-mile long block of middle and lower Ordovician dolomites and limestones with a few Permian Hueco Limestone remnants on top and Proterozoic Van Horn Sandstone at base. About 950 feet of the Ordovician strata are exposed on the peak.

The Salt Basin Rift cuts diagonally across the Diablo ridge in front of the Baylor Mountains and runs up to the Guadalupe Mountains at about 9 o'clock, where on a clear day you can see El Capitan (8,085 feet), capped by Capitan Limestone. There, because of the uplift, the limestone is nearly 3,000 feet higher than on the Apache Mountains (5,216 feet), and 5,800 feet higher than in an oil exploration well, the Plymouth No. 1, drilled 24 miles south of Balmorhea.

The Milwhite talc plant on the right processes talc from Tumbledown Mountain, on the west side of the Beach Mountains north of Van Horn, mainly for wall tile. The company also quarries bentonite south of Alpine.

Take the Van Horn exit off Interstate 10 at Mile 36.0 and turn left on US 90. Reset your odometer: mileages in the next segment are measured from this junction.

## Stop 3: Van Horn

Van Horn, the county seat of Culbertson County, was founded when the Texas and Pacific railroad arrived in 1881 and today is a busy stop on Interstate 10, offering the traveler a variety of motels, restaurants and gas stations.

From Van Horn, Highway 90 runs down the Lobo Valley graben in the Salt Basin Rift between the Wylie Mountains (5,310 feet) on the left and the Carrizo Mountains (5,304 feet) on the right. According to Henry, Gluck and Bockhoven, basin fill in the graben is 4,000 feet thick here and overlies Permian strata that have been down faulted 4,500 feet.

The Carrizo Mountains at 3 o'clock consist of metamorphosed igneous and sedimentary rocks of the Carrizo Mountain Group, 7,700 and 19,000 feet thick respectively. Bedding planes in the face towards the road dip to the left (Fig. 10.9).

The Eagle Mountains at 3 o'clock are on a horst or uplift block between the Eagle Flat graben on the right, which continues across the Rio Grande west of Valentine, and the Red Light graben, running down to the river on the far side of the mountains (Fig. 10.1). The high point in the range is Eagle Peak (7,484 feet), the highest mountain for many miles. It is built up of 3 rhyolite lava units surmounted by an intrusion and a volcanic caldera (Fig. 10.10).

The Van Horn Mountains are on the horizon from 12 to 3 o'clock as you approach the railroad underpass, 7.4 miles south of Van Horn. They are a horst between the Eagle Flat graben on their far side and the Lobo Valley graben on the highway side. The mountains are capped by Cretaceous Cox Sandstone with patches of volcanic rocks on top.

The sandstone dies away to the north and the underlying Permian Hueco Limestone is at the surface as evidenced by the dimpled appearance of the terrain at 3 o'clock.

The mountains also include a caldera, the Van Horn Mountains caldera. High Lonesome Peak (5,825 feet), about 7 miles away at 1 o'clock and the high point in the range, is at its center (Fig. 10.11). The caldera produced two ash-flow tuffs and developed into a depression almost 400 feet deep during eruption of the second one. Subsequently it filled up with air-fall tuff, lavas and the second ash-flow tuff, which were then intruded by a rhyolite dome and a basalt plug.

The Three Sisters buttes at 9 o'clock are outliers of tuffs from the caldera overlain by basalt (Fig. 10.12). Canning Ridge to their right at 9 o'clock (5,194 feet) is a syenite intrusion in what is thought to be another caldera, the Wylie Mountains caldera, stretching back about 4 miles behind the ridge.

At Mile 9.5, the road goes down a low scarp over the boundary fault between the Van Horn Mountains and the Lobo Valley graben (Fig. 10.1). Basin fill, more than 500 feet thick here, broadens out in front as the boundary fault veers off to the right.

## Stop 3: Van Horn Wells (10.6 miles from Van Horn)

The historic marker for the wells is beside the highway; the wells are a short distance back from the road on private land. They were a stopping point on the San Antonio-El Paso route. Major Jefferson Van Horne, on his way to establish what is now Fort Bliss near El Paso, halted here in 1849 with an army train of 275 wagons, not long after the return to San Antonio of Lieutenant Whiting and his party.

Whiting did not see these wells having come downriver from El Paso and up to the Lobo Valley via Needle Peak, 10 miles ahead, but their presence would have been well-marked by Apache trails; at that time, local Apaches wintered along the Rio Grande and summered in the Davis Mountains. Unofficial traffic was heavy on the route. In July 1849, it was estimated that as many as 1,200 wagons and 4,000 emigrants were camped near El Paso, waiting for army permission to leave for California and its gold.

Another Van Horn, James Judson, 2nd Lieutenant, U.S. Army, commanded a company of the Eighth Infantry and, according to the Handbook of Texas, garrisoned the wells from 1859 until he was taken prisoner by Confederate forces in 1860, and it is thought that the wells and the town were named after him.

Fig. 10.9: The Carrizo Mountains at daybreak from Mile 3.0. Thick beds of metamorphosed sedimentary rocks dip away from the road and slightly to the left. The summit of Eagle Peak is at the extreme left.

Fig. 10.10: Eagle Peak (7,484 feet) from Mile 4.7. The mountain, 14 miles west of the highway, is built up of three rhyolite lava flows with an intrusion and caldera at its summit.

Fig. 10.11: In this view of the Van Horn Mountains taken from Mile 8.0, dimpled Permian Hueco Limestone hills are on the right skyline. Volcanic rocks from the Van Horn Mountains caldera make up the hills at front left and continue to High Lonesome Peak on the rear skyline. The caldera was active around 37-38 Ma, significantly before most of the Davis Mountains volcanic centers.

Fig. 10.12: The Three Sisters from Mile 8.0. These are volcanic remnants capped by basalt above trachyte from the Van Horn Mountains caldera. Volcanic tuff from the Eagle Mountains caldera occurs at plain level.

Heliograph Hill (5,035 feet) above the wells is another reminder of the military presence. The heliograph was an instrument used in the nineteenth century by army units to send Morse code signals by sunlight reflected with mirrors. They could be seen by the naked eye at up to 30 miles and much farther by telescope. The high ground at about 1 o'clock from the historical marker is part of a basalt plug in the east flank of the Van Horn Mountains caldera. At 2:30 o'clock you can see level Cox Sandstone strata, here more than 1,000 feet thick.

Just beyond the wells, the sharp peak of Chispa Mountain (5,200 feet) is prominent at 11 o'clock (Fig. 10.13). The peak is capped by trachyte flows overlying volcanic and volcaniclastic rocks. Several similar steep-sided volcanic hills occur to its right. The pecan orchard on the left at Mile 12.4 is irrigated by water pumped from basin fill. Irrigated fields on the right grow mainly hay and alfalfa for cattle feed. The view to the right from Mile 14.1 is given in Fig. 10.14.

The abandoned grocery store, motel and gasoline station at Mile 15.9 is all that is left of Lobo, at one time the center of a flourishing irrigation district. Development began about 1948 when the depth to the water table was about 90 feet. By 1958, 52 wells were producing water and the water table was being lowered by about 3 feet per year. By the late 1960's, irrigation was greatly reduced as water became too deep to be extracted profitably, an excellent example of "water mining". The only recharge is from scant rainfall, most of which never reaches the water table.

At the county line, Mile 19.5, you can look through a pass at 2 o'clock between the Van Horn Mountains (5,565 feet) on the right and the Sierra Vieja (6,450 feet) on the left to high mountains on the skyline in Mexico. The sharp, dark Needle Peak (4,336 feet) is in the middle of the pass.

## Stop 3: Ranch Road 2107 (21.0 miles from Van Horn)

A county road branches off Ranch Road 2107 4 miles from Highway 90, crosses the mountains at Needle Peak and runs down to the Rio Grande. It follows Whiting's return route from El Paso to San Antonio:

> "Following a number of greatly traveled trails, we were led to an opening of the hills, a narrow gate just above the columns, between two bluffs over- and underlaid with red clay and occasional green sand ... A little creek, which cannot be seen until the traveler is upon it, winds its way, destitute of trees or bushes, through its lowest parts. Here we found plenty of water and sufficient grazing for our train. Scarcely any wood, however, is to be found in this place. It is a residence of the Apache, whose lodges are seen here in great numbers especially about the "Needle", a singular rock coming sharp up to a point and apparently a column of basalt".

Henry, Price and Parker identified Needle Peak as a basaltic diatreme containing blocks of conglomerate of probable Tertiary age intruded into Upper Cretaceous rocks. A diatreme is a breccia-filled volcanic pipe formed by a gaseous explosion. Julius Dasch mapped the peak as part of a linear feature, probably a dike.

The flat tops of the Sierra Vieja to the left of the pass are capped by Bracks Rhyolite of similar age and composition to the Star Mountain rhyolite and the Crossen Trachyte. The first mesa to the right of the pass is capped by the Buckshot Ignimbrite, which was erupted from the Infiernito caldera, 50 miles to the south (ignimbrite is another name for a welded or ash-flow tuff). The second mesa to the right is capped by High Lonesome Tuff from the Van Horn Mountains caldera and the third mesa by Bracks Rhyolite.

About Mile 30.3, the highway crosses into deepest graben in the Salt Basin Rift, the Valentine basin, which produces the lowest gravity reading in the rift, -160 milligals, indicating a thick section of lower density rocks (Fig. 10.4). Strata in the basin, as measured by the Killam-Means #1 oil exploration well drilled about a mile north of Valentine, are almost 8,000 feet lower than in the Sierra Vieja. The rift's floor is uneven, however, probably due to block faulting. In the Sinclair #1 Evans well drilled 16 miles south of Valentine, for example, the base of the volcanic rocks is 2,000 feet higher than in the Killam-Means well. Incidentally, the Sinclair well drilled down to the Proterozoic, 9,400 feet below the surface.

The Y-6 Hills at 9 o'clock from Mile 30.5 are capped by basalt over rhyolite. The highway for the next eight miles passes low basalt or basalt-capped hills and ridges on the right near the center of the Valentine basin and was called Rubio Ridge by Wightman. The section in Fig. 9.4 from the Y-6 Hills crosses the highway about this point and continues through Rubio Ridge to the Sierra Vieja.

## Stop 4: Chilicote Ranch Entrance (36.5 miles from Van Horn)

This is a good point to stop and look over the mountains to the left. The high pointed peak at 9 o'clock, Spring Mountain (6,752 feet, Fig. 10.15), is a nepheline trachyte intrusion, one of a cluster surrounding the El Muerto Spring area and part of a long chain of similar intrusions extending from New Mexico through the Davis Mountains to the Big Bend. The long ridge on the left of Spring Mountain is capped by Moore Tuff, erupted from the small El Muerto caldera (see page 32).

Sawtooth Mountain is the craggy peak at 9:30 o'clock with Bear Mountain just in front. At 10 o'clock, Mount Livermore is capped by an eroded trachyte dome complex. Chapter 7 describes these mountains in detail.

Fig. 10.13: Chispa Peak (5,200 feet), seen in silhouette from Mile 10.2, rises 1,150 feet above Highway 90. It and the other peaks nearby are capped by a widespread basalt flow underlain by a set of trachytic tuffs and lavas called the Garren Group, also widespread around this area and of unknown origin.

Fig. 10.14: The view across the Lobo Valley graben to the Sierra Vieja at Mile 14.1 from Van Horn. Dark Cretaceous Cox Sandstone dips under thick bedded light gray Loma Plata Limestone on the left. The limestone occurs from the Van Horn Mountains west and is nearly 700 feet thick here. Cretaceous strata thicken towards the Chihuahua Trough to the west. The upper levels on the right are of Colmena Tuff, the lower of the tuffs produced by the Van Horn Mountains caldera.

Fig. 10.15: Spring Mountain (6,752 feet), the high peak on the right, from Mile 36.5. Brushy Mountain is on the left and Baldy Mountain in front. All are intrusions into Moore Tuff.

Fig. 10.16: Looking down the Wild Horse/Capote Draw graben of the Salt Basin Rift from Mile 39.0. Capote Peak on the right and Cleveland Peak (6,549 feet) to its left are tilted into the graben. Chinati Peak is in the middle of the graben with the Oak Hills to left sloping gently to the east.

The highway enters Valentine at Mile 37.4. Valentine was founded as a watering point on the railroad and became a shipping point for local cattle ranchers. Its post office is popular when it comes time to send Valentine cards.

Dumas, Dorman and Latham located the 1931 epicenter of the largest earthquake by far ever recorded in Texas at about 8.5 miles to the northwest, probably on a fault on the northeast boundary of the Valentine basin. They estimated the magnitude of the earthquake to have been between 5.6 and 6.4 on the Richter scale. Interestingly, the shock was barely felt in the Mexican towns across the Rio Grande although distinctly felt hundreds of miles away in all other directions. This is thought to be due to the presence of thousands of feet of salt and gypsum evaporites in the Rio Grande valley, deposited during the Cretaceous period in the Chihuahua Trough. Evaporites transmit shock waves poorly.

## Stop 5: Leaving Valentine (38.2 miles from Van Horn)

Leaving Valentine, Wild Horse Draw and its extension Capote Draw run down towards Chinati Peak (7,721 feet) at 1:30 o'clock. The Salt Basin Rift splits about 9 miles south of Valentine, one branch going down Wild Horse/Capote where it terminates at FM 2810 near the head of Pinto Canyon, the other branch continuing along Ryan Flat towards Marfa.

The Capote Draw branch is a half-graben, faulted only on the left or east side, with strata on the right tilted into the graben as shown by Capote Peak (6,212 feet, Fig. 10.16) on the right. Capote Peak is capped by Mitchell Mesa Welded Tuff overlying tuff and Bracks Rhyolite. The welded tuff, a widespread, thin and very durable rhyolite produced from the Chinati Mountains caldera, is found as far as 50 miles northeast and 50 miles south of the caldera.

The graben boundary fault runs to the right of Oak Hills (6,063 feet), the gently east-dipping cuesta at 1 o'clock, where it displaces strata 1,000 feet down into the graben.

Behind Oak Hills, Cuesto del Burro (5,879 feet) is another volcanic complex at the site of the Infiernito caldera. This caldera is the oldest so far found in the Trans-Pecos area, dated at about 37.3 Ma.

## Stop 4: Junction of Highway 90 and Ranch Road 505 (44.7 miles from Van Horn)

At this point, you can either turn left on to Ranch Road 505, the Valentine Cutoff, which connects with Highway 166, the Scenic Loop around the western Davis Mountains, and then to Highway 17 and Fort Davis.

Alternatively, you can continue on Highway 90 to Marfa, as described below and turn left on to Highway 17 to Fort Davis.

You get a very clear picture of the Davis Mountains from about 9:30 to 12:30 o'clock. The sharp peak at 12 o'clock is Blue Sheep Mountain (see page 112). Along both sides of the road for several miles, the road crosses level grassy meadowland.

At Mile 49.8, the road climbs over a low ridge and bears slightly left. The butte with the sharp-pointed peak at 9 o'clock is El Muerto Peak, an intrusion. Basalt crops out on a ridge on the right.

At the junction with Highway 166 (Mile 53.8), turn right towards Fort Davis, 23 miles ahead. The description forward continues on page 112.

## Junction of Highways 90/505 to Marfa

Some may prefer to drive back to Fort Davis via Marfa rather than take the Valentine Cutoff. This short road description is for them.

Two and a half miles ahead, the low cuestas about 5 miles away on the right at Mile 47.5 are capped by basalt.

The ridges from 10 to 11 o'clock at Mile 50.9 are mostly ash-flow tuffs of the Barrel Springs Formation. Mount Livermore and the high bare cliffs of Brooks Mountain are in full view from the picnic area at Mile 51.1.

The Air Force tethered aerostat radar site (TARS) at Mile 52.5 operates balloon-borne radar to provide low level surveillance for drug interdiction.

At the top of a rise at Mile 64.5, the mountains of the Alpine area are in view with the twin rhyolite intrusions of the Haystacks at 11 o'clock and the Puertacitas Mountains to their left. Cathedral Mountain is just to the left of the highway on the skyline with the whale-backed ridge of Cienega Mountain and then Goat Mountain to its right (see page 141 for descriptions of these mountains).

Coming into Marfa, Ranch Road 2810 on the right leads to Ruidoso on the Rio Grande. The road is good-quality hardtop for the 35 miles to the top of Pinto Canyon. Beyond there, the gravel road down the canyon can get washed out by rain and it is best to ask about its condition at the Presidio County offices before using it in the summer or fall.

The junction of US 90 with US 67 south to Presidio and Chihuahua and TX 17 north to Fort Davis is at Mile 72.5. Turn left on to Highway 17. The description of this section of road begins on page 145.

# Glossary

Words in italics are defined elsewhere in the glossary.

**alkali:** Said of *igneous rock* with more sodium and or potassium than is required to form *feldspar* with the available silica.

**alkali feldspar:** A group of minerals containing the *alkali metals* sodium and or potassium but little calcium.

**alkali metal:** Any of the elements lithium, potassium, sodium, rubidium or cesium.

**alluvial fan:** A low gently-sloping mass of loose *alluvium*, shaped like a fan, left by a stream at the point where it comes out of a narrow mountain valley into a plain or broad valley.

**alluvium:** Sand, clay, silt or gravel deposited recently and unconsolidated, i.e. not cemented together.

**anticline:** A fold, generally convex up, whose core contains older rocks.

**ash-fall tuff:** Tuff created by airborne volcanic ash falling from an ash cloud erupted from a volcano.

**ash-flow tuff:** Tuff created from an ash flow, a mixture of volcanic gases and particles, usually hot, that flows out from explosive viscous magma in a volcanic fissure or crater; synonymous with *pyroclastic* flow.

**asthenosphere:** A weak layer in the Earth's upper mantle, underlying the lithosphere, in which the amplitude of seismic waves is strongly reduced. It is generally thought to contain a small amount of silicate melt and to be the source of magma.

**autobreccia:** A breccia in which parts of a lava crust are incorporated in the still-fluid part.

**basalt:** A dark-colored *igneous rock* composed mainly of calcic plagioclase and pyroxene.

**breccia:** A coarse-grained rock made up of angular broken rock fragments held together by mineral cement or in a fine-grained matrix. A breccia differs from a *conglomerate* in that the fragments have sharp edges and unworn corners.

# GLOSSARY

**caldera:** A large basin-shaped volcanic depression. A collapse caldera is created by the collapse of a *magma chamber* through the removal of magma by volcanic explosions or lava eruptions, or by the removal of magma through subterranean pathways.

**calcite:** The principal component of limestone, calcium carbonate $CaCO_3$.

**claystone:** A weakly *indurated, sedimentary rock* made up of mainly clay particles i.e. those with diameters of less than 0.01 millimeter.

**conglomerate:** A coarse-grained *sedimentary rock*, composed of rounded or sub-angular fragments of rock larger than 2 mm in diameter, set in a fine-grained matrix of sand or silt and commonly cemented by calcium carbonate, iron oxides, silica or hardened clay.

**cordillera:** This Spanish word for a chain or range of mountains used by geologists to refer a mountain province, especially the great mountain region of the United States from the east face of the Rocky Mountains to the Pacific Ocean

**craton:** A stable part of the Earth's crust, unchanged for many millions of years; usually of Proterozoic age i.e. more than 550 million years old.

**cryptocrystalline:** A rock whose crystals are too small to be recognized under an ordinary microscope.

**crystalline:** A rock composed of crystals that contains no glass.

**cuesta:** A hill or ridge with a gentle slope conforming to the bed or beds that form it on one side and a steep slope on the other formed by outcrops of resistant rocks, the formation of the ridge being controlled by the *differential erosion* of the gently inclined strata.

**devitrified:** Said of volcanic material that has converted from glass to crystalline material.

**dike:** A broad, tabular *igneous intrusion* that cuts across the bedding or foliation of the rock into which it was intruded.

**differential erosion:** Erosion that occurs at varying rates caused by differences in the hardness or resistance of rocks; softer or weaker rocks are eroded more quickly than harder or more resistant rocks.

**evaporite:** A rock composed of minerals that crystallized from salt water or other solutions as they evaporated. Evaporites include gypsum, rock salt, dolomite and various nitrates and borates.

**exfoliation:** The process by which concentric scales or shells of rock are stripped from the surface of a large rock mass.

**extrusive rock:** An *igneous rock* formed from magma that has erupted onto the surface of the Earth; includes *lavas, pyroclastic* flows and volcanic ash.

**fault scarp:** A steep slope or cliff formed by movement along a fault and corresponding to the exposed surface of the fault before it was modified by weathering or erosion.

**feldspar:** A mineral of the *alkali* aluminum silicate feldspar group, the most widespread on Earth, making up 60 per cent of the Earth's crust.

**felsic:** Said of an *igneous rock* rich in feldspars and silica, a light-colored rock. It is the complement of mafic.

**graben:** A trough or basin bounded on both sides by normal faults dipping into the graben which has moved down relative to the adjoining fault blocks. See also *horst*.

**groundmass:** The material between phenocrysts in a porphyritic igneous rock. It is finer grained than the phenocrysts and may be crystalline, glassy or both.

**half graben:** A basin bounded on one side by a normal fault.

**horst:** A block of the Earth's crust that is bounded on opposite sides by faults dipping away from the block and has moved upward relative to the two adjoining blocks. See also *graben*.

**igneous rock:** A rock made from molten or partly molten material, i.e. magma, that has cooled and solidified.

**indurated:** Said of a rock hardened or consolidated by pressure, cementation or heat.

**intrusion:** A rock that has formed from a magma that has intruded into pre-existing rock.

**isotopic dating:** A method by which the ratio of certain isotopes in an igneous rock is calculated to give the age of the rock. The most modern method, which measures the ratios of two argon isotopes, is accurate to within 0.3%.

**laccolith:** An *igneous intrusion* parallel to the bedding into which it was intruded except for its roof which is domed.

**lahar:** A mudflow, consisting mainly of volcanic rocks and debris, on the flanks of a volcano.

**lava:** *Magma* that comes to the Earth's surface through a volcanic vent or fissure.

**lithosphere:** The rigid outer layer of the Earth's surface, made up of the crust and the upper mantle, typically 22 to 125 miles thick.

**mafic:** Said of an *igneous rock* that is composed of minerals rich in iron and magnesium, a dark-colored rock. It is the complement of felsic.

**magma:** Naturally occurring mobile rock material, generated within the Earth and capable of being extruded and intruded, from which igneous rocks are derived through cooling. It may or may not contain suspended solids such as crystals and rock fragments.

**metamorphic:** Said of a rock that has been modified from another by heat, pressure or chemical influences, usually at depth in the Earth's crust.

# GLOSSARY

**nepheline:** Alkali silicate of the feldspathoid group (Na,K)[AlSiO$_4$].

**nepheline syenite:** A *plutonic* rock composed of *alkali feldspar* and *nepheline* and perhaps *alkali mafic* minerals, the coarse-grained equivalent of phonolite. Nepheline syenite or phonolite intrusions are found in the Trans-Pecos along a zone from the Black Hills east of Persimmon Gap to the Cornubas Mountains in New Mexico, including Santiago Peak, Black Mesa, Elephant Mountain and Paisano Peak.

**normal fault:** A fault in which the hanging or upper wall has dropped relative to the foot or lower wall.

**orogeny:** The process by which structures in fold-belt areas were formed, including thrusting, folding and faulting. In the Cenozoic era, such folding and faulting led to the formation of mountainous landscapes but in earlier eras, fold belts are seldom associated with mountainous terrain.

**outlier:** An area or group of rocks surrounded by older rocks.

**palisades:** Jointed cliffs in which erosion around the jointing has created columns running from top to bottom. The name comes from the Palisades, a line of cliffs along the Hudson River in New York and New Jersey.

**pediment:** A gently-sloping broad erosion surface in a semi-arid region at the base of an abrupt and receding mountain front or plateau escarpment.

**phenocryst:** A relatively large crystal in a *porphyritic* rock.

**plug:** A vertical igneous pipe, the channel by which magma reached a volcanic vent.

**pluton:** An *igneous intrusion* formed at depth, of area greater than 40 square miles and with no known floor.

**porphyritic:** Describes a rock in which larger crystals are set in a finer grained groundmass.

**pyroclastic flow:** A flow of *pyroclasts*, usually very hot; synonymous with ash flow.

**pyroclastic surge deposit:** A stratified layer of *pyroclasts* left by a cloud of hot particles and gas moving turbulently along the ground from a volcanic vent.

**pyroclast:** An individual particle of rock ejected by an explosive volcanic eruption; from the Ancient Greek words for fire and broken into pieces; the adjective is pyroclastic.

**rheomorphic:** Said of a rock that became mobile through at least partial fusion; also the process producing such a rock.

**rhyolite:** A group of fine-grained extrusive rocks, typically *porphyritic* and commonly exhibiting flow structures, with *phenocrysts* of quartz and *alkal feldspar* in a glassy to *cryptocrystalline* groundmass. Rhyolite grades into *trachyte* with a decrease in quartz content.

**sedimentary rock:** A rock that has formed from the consolidation of loose sediment such as fragments of older rock, chemically precipitated material, volcanic pyroclastic fragments, and organic remains.

**schist:** a *metamorphic* rock in which crystals of platy minerals such as mica give it a strong foliation.

**sill:** A broad, tabular *igneous intrusion* that parallels the bedding or foliation of the rock into which it was intruded.

**stock:** An *igneous intrusion* formed at depth of area 40 square miles or less, roughly circular in outline and with no known floor.

**syenite:** A group of plutonic rocks containing *alkali feldspar*, plagioclase, one or more mafic minerals, with quartz, if present, only as an accessory; the intrusive equivalent of trachyte; with increasing quartz, grades into granite.

**syncline:** A fold, generally concave up, whose core contains younger rocks.

**thrust fault:** A fault where the upper or hanging wall has moved over the lower or foot wall, shortening the Earth's crust.

**trachyte:** A group of fine-grained extrusive rocks, generally *porphyritic*, having *alkali feldspar* and minor *mafic* minerals as the main components. the extrusive equivalent of syenite. Trachyte grades into *rhyolite* as the quartz content increases.

**tuff:** A rock composed of consolidated or cemented volcanic ash; includes *ash-flow tuff* and *ash-fall tuff*.

**unconformity:** A substantial break or gap in the geological record where a rock unit is overlain by one that is not next in stratigraphic succession; usually the result of a period in which strata were not deposited or where intervening strata were eroded before the upper unit was deposited.

**vitrophyre:** a *porphyritic igneous rock* with a glassy *groundmass*.

**volcanic breccia:** A *volcaniclastic* rock composed mainly of volcanic fragments greater than 2 mm in diameter.

**volcanic dome:** A steep-sided, rounded extrusion of very viscous lava squeezed out of a volcanic vent and forming a mass above the vent.

**volcanic vent:** An opening in the Earth's surface through which volcanic materials are extruded.

**volcaniclastic:** Pertaining to a sedimentary rock composed of mainly or partly broken fragments of volcanic origin.

**welded tuff:** A rock composed of *pyroclasts* welded together by a combination of heat of the particles, the weight of overlying material and hot gases.

# Reading List

Anderson, J.E. Jr., 1968, Igneous geology of the central Davis Mountains, Jeff Davis County, Texas. The University of Texas at Austin, Bureau of Economic Geology Quadrangle Map 36, scale 1:62,500, 1 sheet, 18 p. text.

Baldridge, W. Scott, 2004, *Geology of the American Southwest*: Cambridge University Press, 280 pp.

Barnes V.E., Project Director, 1982, Fort Stockton Sheet, in Geological Atlas of Texas. *Tex. Bur. Econ. Geol.*, scale 1:250,000.

Bridwell, R.J., 1978, Physical behavior of upper mantle beneath northern Rio Grande Rift in Hawley, J.W., ed., Guidebook to Rio Grande Rift in New Mexico and Colorado. New Mexico Bureau of Mines and Mineral Resources, Socorro, NM, p. 228-230.

Chapin, C.E., and S. M. Cather, 1994, Tectonic setting of the axial basins of the northern and central Rio Grande Rift. *In* Keller, G.R. and Cather, S.M., eds., Basins of the Rio Grande Rift: Structure, stratigraphy and tectonic setting. *Geol. Soc. Am. Spec. Paper* **291**, p.5-26.

Chowdury, A.H., Ridgeway, C., and R.E. Mace, 2004, Origin of the waters in the San Solomon Spring system, Trans-Pecos Texas. *In* Aquifers of the Edwards Plateau. Texas Water Development Board, Groundwater Report 360, 30 p.

Collinsworth, B.C., and D.M. Rohr, 1986, An Eocene carbonate lacustrine deposit, Brewster County, West Texas. *In* Pause, P.H. and Spears, R. G., eds., Geology of the Big Bend and Solitario Dome: *West Texas Geol. Soc. 1986 Field Trip Guidebook*, p.117-24.

Daugherty, F.W., and P.W. Dickerson, 1980, Supplementary road log no.3 Van Horn to Alpine. *In* Dickerson, P.W. and Hoffer, J.M., eds. Trans-Pecos Region Southeastern New Mexico and West Texas: New Mexico Geological Society, thirty-first field conference, p. 48-52.

Dasch, E.J., Armstrong, R.L. and S.T. Clabaugh, 1969, Age of the Rim Rock dike swarm. *Geol. Soc. Am. Bull.* **80**, 1819-1824.

Dickerson, P.W., and W.R. Muehlberger, 1994, Basins in the Big Bend segment, Rio Grande Rift. *In* Keller G.R., and Cather S.M., eds., Basins of the Rio Grande Rift: structure, stratigraphy, and tectonic setting. *Geol. Soc. Am. Spec. Paper* **291**, 283-297.

Dumas, D.B., Dorman, H.J., and G.V. Latham, 1980, A reevaluation of the August 16, 1931 Texas earthquake. *Bull. Seismological Soc. Am.* **70**, 1171-1180.

Eifler, G.K., 1951, Geology of the Barrilla Mountains, Texas. *Geol. Soc. Am. Bull.* **62**, 339-353.

Elkins, R.G., 1986, Laccoliths of the Musquiz Canyon area southern Davis Mountains, Trans-Pecos Texas. Bachelor of Science thesis: Waco, Texas, Baylor University, 111 p.

Gao, W., Grande S.P., Baldridge W.S., Wilson D., M. West, Ni J.F., and R. Aster (2004), Upper mantle convection beneath the central Rio Grande rift imaged by P and S wave tomography. *J. Geophys. Res.* **109**.

Goetz, L.K., 1985, Salt Basin Graben: A basin and range right-lateral transtensional fault zone – some speculations. *In* Dickerson, P.W. and Muehlberger, W.R, eds., Structure and tectonics of Trans-Pecos Texas: West Tex. Geol. Soc. **85-81**, p.165-168.

Gorski, D., 1970, Geology and trace transition element variation of the Mitre Peak area, Trans-Pecos Texas [M.A. Thesis]. Austin, Texas, University of Texas at Austin, 201 p.

Hempkins, W.B., 1962, Geology and petrography of the Sawtooth Mountains area, Jeff Davis County, Texas. Unpublished M.A. thesis, University of Texas at Austin, 144 p.

Henderson, G.D., 1989, Geology of the Medley kaolin deposits and associated volcanic rocks, Jeff Davis County, Texas [M.S. Thesis]. Waco, Texas, Baylor University, 99 p.

Henry, C.D., 1998, Geology of Big Bend Ranch State Park, Texas. *Geol. Soc. Am. Bull.* Guidebook No. 27, 72 p.

Henry, C.D., M.J. Kunk, and W.C. McIntosh, 1994, 40Ar/39Ar chronology and volcanology of silicic volcanism in the Davis Mountains, Trans-Pecos Texas: Geological Society of America Bulletin v.106 p.1359-76.

Henry, C.D., J.K. Gluck, and N.T. Bockoven, 1985, Tectonic map of the Basin and Range Province of Texas and adjacent Mexico: The University of Texas at Austin, Bureau of Economic Geology Miscellaneous Map No. 36.

Henry, C.D., and J.G. Price, 1989, Volcanic rocks of the Ft. Davis area, Davis Mountains in Muehlberger, W.R. and P.W. Dickerson, eds., Structure and Stratigraphy of Trans-Pecos Texas: 28th International Geological Congress, Field Trip Guidebook T317 p. 141-146.

Henry, C.D., J.G. Price and D.F. Parker, 1989, Alternate road log from Van Horn to Alpine in Price, J.G., C.D. Henry, D.F. Parker and D.S. Barker, eds., Igneous geology of Trans-Pecos Texas. *Tex. Bur. Econ. Geol.*, Guidebook No. 23, p. 91-94.

Henry, C.D., J.G. Price and D.F. Parker and J.A. Wolff, 1989, Mid-Tertiary silicic alkalic magmatism of Trans-Pecos Texas: Rheomorphic tuffs and extensive silicic lavas *in* Chapin, C.E and J. Zidek, eds., Field excursions to volcanic terranes *in* the western United States, Volume I: Southern Rocky Mountain region; New Mexico Bureau of Mines and Mineral Resources, Memoir 46, p.231-274.

Hill, C.A., 1999, Reevaluation of the Hovey Channel in the Delaware Basin, West Texas: AAPG Bulletin, **83**, p.277-294.

Keller, G.R., and W.J. Peeples, 1985, Regional gravity and aeromagnetic anomalies in West Texas *in* Dickerson, P.W. and W.R. Muehlberger, eds., Structure and tectonics of Trans-Pecos Texas: West Tex. Geol. Soc. **85-81**, p.101-106.

King, P.B., and P.T. Flawn, 1953, *Geology and mineral deposits of Pre-Cambrian rocks of the Van Horn area, Texas.* Tex. Bur. Econ. Geol., 218p.

Mattison. G.D., 1979, A reinterpretation of the Sheep Pasture tuffs. *In* Walton, A.W. and Henry C.D., eds., Cenozoic geology of the Trans-Pecos volcanic field of Texas. *Tex. Bur. Econ. Geol.* Guidebook **19**, p.83-91.

McAnulty, W.N., 1950, Geology and ground-water resources of Alpine and adjacent territory, Brewster County, Texas: Unpublished manuscript.

Nelson, D.O., and K.L. Nelson, 1986, Supplementary road log – Alpine-Fort Davis-Davis mountain loop in Pause, P.H. and R.G. Spears, eds., Geology of the Big Bend and Solitario Dome: *W. Tex. Geol. Soc.* 1986 Field Trip Guidebook, pp. 63-73.

Parker, D.F., and F.W. McDowell, 1979, K-Ar geochronology of Oligocene volcanic rocks, Davis and Barilla Mountains, Texas: Geol. Soc. Am. Bull **90**, p.1100-10.

Parker, D.F., 1979a, The Paisano Volcano: Stratigraphy, age, and petrogenesis. *In* Walton, A.W., and Henry, C.D., eds., Cenozoic geology of the Trans-Pecos volcanic field of Texas. *Tex. Bur. Econ. Geol.* Guidebook 19, p. 97-105.

Parker, D.F., 1979b, The Paisano volcano: Field trip stops in Walton, A.W., and Henry, C.D., eds., Cenozoic geology of the Trans-Pecos volcanic field of Texas: *Tex. Bur. Econ. Geol.* Guidebook 19, 193 p.

Parker, D.F., 1989, Igneous geology of the Davis Mountains, West Texas in Muehlberger, W.R. and P.W. Dickerson, eds., Structure and Stratigraphy of Trans-Pecos Texas: 28th International Geological Congress, Field Trip Guidebook **T317** p. 135-140.

Parker, D.F., and L. Gilmore, 1991, Road Logs. *In* Parker, D.F., Tsuchia, L.C. and C.L. McKnight, eds., The Davis Mountains volcanic field: Baylor Geological Society.

Pearson, B.T., 1981, Some structural problems of the Marfa Basin area, in Johns, R.D ed., Marathon-Marfa region of West Texas: Soc. Econ. Paleontologists and Mineralogists, Permian Basin Section **81-20**, p.59-73.

Pearson, B.T., 1985, Tertiary structural trends along the northeast flank of the Davis Mountains. In Dickerson, P.W. and W.R. Muehlberger, eds., *Structure and tectonics of Trans-Pecos Texas*. West Texas Geological Society **85-81**, p.153-157.

Price, J.G., Henry C.D., Parker D.F., and D.S. Barker, 1986 eds. Igneous geology of Trans-Pecos Texas: *Tex. Bur. Econ. Geol. Guidebook* **23**, 360 p.

Reilinger, R.E., Brown, L.D., and J.E. Oliver, 1979, Recent vertical crustal movements from leveling observations in the vicinity of the Rio Grande rift. *In* Riecker, R.R., ed., *Rio Grande Rift: Tectonics and Magmatism*: Am. Geophysical Union, Washington, D.C., p.223-236.

Rodríguez, José Policarpo, *The Old Guide: His Life in His Own Words*. Dallas: The Methodist Episcopal Church, 1897.

Seager, W.R., and P. Morgan, 1979, Rio Grande Rift in southern New Mexico, West Texas and northern Chihuahua *in* Riecker, R.R., ed., *Rio Grande Rift: Tectonics and Magmatism*. American Geophysical Union, Washington, D.C., p.87-106.

Smithers, W.D., 2000, *Chronicles of the Big Bend: A Photographic Memoir of Life on the Border,* College Station, Texas A & M University Press.

Spearing, D., 1991, *Roadside Geology of Texas*. Mountain Press Publishing Company, Missoula, Montana, 418 p.

Twiss, P.C., 1979, Marfa sheet. *Tex. Bur. Econ. Geol.* Geologic Atlas of Texas, scale 1:250,000

Velduis, J.H., and G.R. Keller, 1980, An integrated geological and geophysical study of the Salt Basin graben, West Texas. *In* Dickerson, P.W. and Hoffer, J.M., eds. *Trans-Pecos Region Southeastern New Mexico and West Texas*: New Mexico Geological Society, thirty-first field conference, p. 141-150.

Whiting, W.H.C., 1849, Journal of William H.C. Whiting. In Bieber, R.P. ed., *Exploring Southwest Trails 1846-1854*: Philadelphia, Porcupine Press, 1974, p.241-350.

Wilson, D., Aster R., Ni J.F., Grand S.P., West M., Gao W., Baldridge W.S., and S. Semken, 2005, Imaging the seismic structure of the crust and upper mantle beneath the Great Plains, Rio Grande Rift, and Colorado Plateau using receiver functions. *J. Geophys. Res.* **110**.

# Index

Adobe Canyon, 74, 98, 99, 100, 104, 113

Adobe Canyon Formation, 27, 29, 35, 36, 53, 54, 76, 100, 103, 104, 113, 121

Alamito Creek, 68, 121, 162, 175

Alpine, 12, 23, 93, 140, 152

Alpine basin, 136, 143, 148, 149, 150, 152, 153, 157

Alvarado Ridge, 168, 171

Antelope Peak, 146

Apache Mountains, 23, 41, 44, 49, 108, 180, 183, 187, 191

Arabella Mountain, 85, 93

autobreccia, 80, 81, 82, 94, 158, 204

Baldy Peak, 118, 125

Balmorhea, 45, 110

Balmorhea Lake, 45

Balmorhea State Park, 45, 47, 49, 112

Barbaras Point, 63

Barillos Dome, 36, 143, 145, 147, 149

Barrel Spring, 127, 132

Barrel Springs Formation, 27, 33, 34, 36, 52, 53, 54, 55, 59, 60, 62, 63, 67, 68, 76, 77, 79, 80, 81, 82, 84, 85, 86, 87, 89, 90, 93, 94, 95, 96, 97, 98, 99, 114, 118, 123, 127, 134, 135, 136, 140, 141, 165, 166, 202

Barrel Springs Ranch, 114, 127, 128

Barrilla Mountains, 41, 43, 44, 45, 46, 52, 55, 57, 111, 112, 210

Basin and Range extension, 57, 168, 173

Basin and Range province, 167, 168, 173

Baylor Mountains, 23, 188, 190, 191

Beach Mountains, 190, 191

Bear Mountain, 100, 118, 119, 123, 198

Belding, 43, 44

Big Aguja Mountain, 55, 58

Big Bend Ranch State Park, 177, 210

Bird Mountain, 150

Bloys Camp Meeting Ground, 130, 131

Blue Mountain, 37, 67, 68, 77, 78, 80, 82, 84, 85, 86, 91, 92, 94, 134, 135, 141, 162, 163, 177

Blue Sheep Mountain, 126, 202

Bone Spring Formation, 182, 183, 188

Boquillas Formation, 53, 54, 103, 104, 109, 176, 182

Boracho Formation, 42, 105, 108, 109, 176, 180, 182, 183

Boracho Peak, 183

Bracks Rhyolite, 32, 197, 201

Brooks Mountain, 38, 113, 118, 119, 120, 121, 126, 162, 202

Brooks Mountain volcanic dome, 32, 34, 101

# INDEX

Brooks Mountain-Mount Livermore dome, 8, 32, 34, 77

Brown Mountain, 91, 94, 129, 131, 132, 162, 163

Buckhorn Caldera, 27, 54, 76, 109, 140, 154

Buda Limestone, 42, 104, 105, 108, 109, 111, 182, 183

Canning Ridge, 193

Capitan Reef, 23, 44, 184

Capote Draw, 200, 201

Capote Peak, 123, 200, 201

Carrizo Mountain Group, 22, 192

Carrizo Mountains, 22, 188, 189, 192, 194

Casket Mountain, 37, 81

Casket Mountain lavas, 26, 35, 37, 76, 81, 113, 134, 135, 140

Castle Rock, 149

Cathedral Mountain, 93, 132, 159, 160, 162, 202

Chihuahua Trough, 23, 172, 188, 201

Chihuahuan Desert Research Institute, 141

Chinati Mountains volcanic center, 12, 35, 162, 167

Chispa Mountain, 196

Cienega Creek, 141, 166

Colmena Tuff, 199

Cottonwood Springs Formation, 27, 30, 35, 148, 151

Cox Sandstone, 108, 122, 176, 182, 184, 192, 196, 199

Crenshaw Mountain, 154

Crossen Trachyte, 27, 29, 32, 35, 140, 141, 149, 150, 151, 197

Crows Nest Hill, 114, 128, 130

Cuesto del Burro, 201

Davis Mountains, 7, 11, 40, 112, 175

Davis Mountains area, 15, 22, 167

Davis Mountains Preserve, 74, 94, 95

Davis Mountains State Park, 11, 74, 75, 76, 86

Davis Mountains volcanic field, 12

Davis Mountains volcanic rocks, 14, 17, 28, 44

Decie Formation, 27, 30, 35, 36, 38, 140, 148, 150, 151, 153, 154, 156

Del Norte Mountains, 150

Del Rio Clay, 42

Delaware Basin, 23, 58, 112, 183, 187, 188

Diablo Platform, 23, 112, 188, 189

Diablo ridge, 188, 191

Dolores Mountain, 136, 166

Eagle Flat, 192

Eagle Mountains, 182, 183, 192, 196

Eagle Peak, 183, 192, 194

East Pacific Rise, 18, 167

El Capitan, 180, 191

El Muerto caldera, 27, 36, 114, 123, 198

El Muerto Peak, 122, 124, 202

El Muerto Spring, 132

Elam Mountain, 150

Elbow Canyon, 94

Elephant Mountain, 12

Encantada Peak, 102, 104

Farallon plate, 18, 19, 20, 25, 167, 168

Finlay Limestone, 122, 180, 182

Fisher Hill, 94

Flattop Mountain, 98, 99, 118

Forbidden Mountain, 55

# INDEX

Fort Davis, 40, 45, 49, 50, 67, 71, 74, 90, 112, 136, 163, 202

Fort Davis National Historic Site, 68

Fort Stockton, 42, 43, 59, 209

Franklin Mountains, 188, 189

Frazier Canyon Creek, 67

Frazier Canyon Formation, 27, 30, 35, 52, 53, 54, 60, 62, 67, 68, 74, 76, 99, 114, 126, 134, 144, 145, 146, 149, 151

Fredericksburg Group, 42, 44

Garren Group, 198

Gavina Ridge, 91, 92, 94

Geronimo Mountain, 116, 118

Glass Mountains, 23, 44, 92, 150, 180

Glen Rose Formation, 41, 42

Goat Canyon Formation, 26, 35, 37, 87, 113, 126, 129, 131, 132, 133

Goat Mountain, 93, 132, 162, 202

Gomez Peak, 48, 53, 104, 107, 108, 109, 110, 111

Gomez Tuff, 27, 29, 30, 32, 35, 46, 52, 54, 58, 60, 62, 63, 67, 76, 103, 104, 105, 108, 109, 111, 137, 140, 144

Grenville Orogeny, 22

Guadalupe Mountains, 23, 172, 180, 183, 184, 191

Guide Peak, 91

H O Canyon, 119, 121

H O Hill, 118, 120, 121

Hancock Hill, 143, 150, 152

Heliograph Hill, 196

Henderson Mesa, 145, 149

Henry Skillman, 127, 131

High Lonesome Peak, 192, 195

High Point, 126

Horse Camp Peak, 102, 104, 183

Horse Mountain, 150

Hospital Canyon, 67, 68, 80

Hueco Limestone, 182, 191, 192, 195

Huelster Formation, 28, 31, 32, 54, 66, 76, 104, 137, 140

kaolin, 130, 131, 210

Keesey Canyon, 77, 78, 80

Kent, 12, 13, 105, 108

La Ristra, 169, 170, 171, 172

Laramide Orogeny, 26, 31, 167

Last Chance Mesa, 32, 145, 149

Limpia Canyon, 11, 62, 66, 67, 75, 78, 81, 85, 86, 105

Limpia Creek, 59, 62, 63, 66, 67, 74, 81, 82, 85, 86, 88, 91, 127, 132, 141, 166, 175, 177

Limpia Crossing, 82, 84, 85

Limpia Formation, 27, 30, 35, 54, 67, 140, 141, 144

Limpia Mountain, 82, 85, 86, 88

Little Aguja Canyon, 91

Little Sawtooth Mountain, 117

Lizard Mountain, 153

Lobo Valley graben, 192, 193, 197, 199

Loma Plata Limestone, 199

Lonely Lee, 122, 124

Lower Cretaceous, 20, 41, 42, 109, 122, 182

Madera Canyon, 54, 95, 96, 111, 177

Mano Prieto, 136, 165

Marfa, 153, 163, 202, 203

Marfa Lights Viewing Center, 159

Marfa plain, 135, 136, 158, 160

McCutcheon Fault, 63, 66

McDaniel Mountain, 96, 97, 117

McDonald Observatory, 11, 74, 85, 90

Merrill Formation, 114, 123, 126

Michigan Flat, 184, 188

Mine Mountain, 130

Mitchell Mesa Welded Tuff, 159, 162, 201

Mitre Peak, 26, 32, 36, 37, 38, 92, 93, 136, 145, 149, 151, 210

Moore Tuff, 27, 36, 114, 118, 123, 124, 198, 200

Mount Fowlkes, 85, 86, 90, 166

Mount Livermore, 38, 82, 85, 95, 97, 112, 119, 120, 121, 126, 132, 162, 177, 188, 202

Mount Livermore volcanic dome, 36, 38, 76, 86, 113, 125, 198

Mount Livermore-Brooks Mountain dome, 34, 177

Mount Locke, 85, 86, 90, 93, 166

Mount Locke Formation, 26, 34, 37, 76, 81, 86, 90, 91, 92, 94, 95, 96, 114, 123, 133, 135

Mount Ord, 92, 93, 150

Musquiz Canyon, 26, 27, 32, 36, 37, 38, 114, 136, 139, 140, 210

Musquiz Creek, 137, 139, 144, 146, 149

Musquiz Dome, 136, 137, 144

Needle Peak, 193, 196, 197

Newman Peak, 108

Nunn Hill, 98, 99

Oak Hills, 200, 201

paisanite, 154

Paisano Baptist Encampment, 155, 158

Paisano Pass, 158

Paisano Pass caldera, 154, 155, 158, 178

Paisano Peak, 35, 154, 155, 159

Paisano Rhyolite, 140, 154, 156

Paisano volcano, 30, 150, 152, 153, 159, 211

Paradise Mountain, 82, 86, 88, 127, 132, 133, 141, 162, 164, 177

Paradise Mountain caldera, 34, 37, 38, 86, 90, 113, 114, 126, 127, 130, 131, 132
inner collapse zone, 130, 132

Pecos, 40, 45, 171

Pecos River, 40, 41, 42, 45, 119, 132, 145, 149, 175

Pecos River drainage basin, 141

Perdiz Conglomerate, 158

Permian Basin, 212

Pine Peak, 37, 82, 85, 95, 177

Point of Rocks, 38, 132, 133, 135

Polks Peak, 145, 148, 149, 150

Pollard dome, 143, 145

Proterozoic, 20, 22, 150, 187, 188, 189, 191, 197, 205

Pruett Formation, 27, 32, 140, 143

Puertacitas Formation, 37, 164

Puertacitas Mountains, 38, 93, 135, 136, 159, 163, 164, 167, 202

Ranger Peak, 151, 153

Red Light graben, 192

Rio Grande, 209

Rio Grande Rift, 168, 169, 170, 172, 188, 189, 209, 212, 213

Robbers Roost Mountain, 96, 98

Rockpile, 116, 117, 118, 119, 120

Round Mountain, 100, 102, 104

Rounsaville Syncline, 52, 57, 59, 111

Rubio Ridge, 171, 176, 177, 197

Ryan Flat, 121, 122, 123, 201

Saddleback Mountain, 46, 55

Salcida Canyon, 91

Salt Basin, 49, 174, 187

Salt Basin Rift, 12, 122, 167, 172, 184, 188, 201

San Antonio Pass, 123

# INDEX

San Solomon Springs, 45, 46, 49

Sawtooth Mountain, 4, 38, 99, 113, 116, 117, 118, 119, 120, 124, 198

Scenic Loop, 74, 112, 166

Scobee Mountain, 67

Sevier Orogeny, 25, 26, 167

Sheep Pasture Formation, 27, 33, 34, 36, 86, 96, 97, 98, 99, 113, 114, 116, 118, 123, 124, 211

Sheep Pasture Mountain, 27, 36, 86, 113, 114, 116, 117

Sierra Diablo Mountains, 183

Sierra Vieja, 12, 32, 122, 123, 162, 171, 177, 196, 197, 199

silicified lava, 127, 128, 129, 131

Skillman's Grove, 130, 131

Skyline Drive, 76, 77, 79, 94

Sleeping Lion Formation, 27, 33, 35, 36, 37, 54, 67, 68, 76, 81, 83, 140, 141, 144, 145

Spring Mountain, 91, 122, 124, 197, 200

Star Mountain, 43, 47, 52, 53, 61, 62, 63, 110, 141

Star Mountain anticline, 59, 108, 173, 174

Star Mountain Formation, 27, 29, 32, 44, 46, 52, 53, 54, 58, 60, 61, 62, 63, 66, 67, 100, 137, 140, 141, 143, 144, 145, 149, 151, 197

Sul Ross State University, 150, 152, 153

Sunny Glen, 142, 148, 151, 152, 153, 154, 157

The Haystacks, 38, 93, 136, 141, 149, 159, 162, 163, 164, 202

Three Points, 96

Three Sisters, 193, 195

Timber Mountain, 53, 55, 111

Tobosa Basin, 23

Toyah Basin, 40, 41, 45

Toyah Creek, 45, 46

Toyahvale, 50, 105, 110

Trans-Pecos Region, 42, 212

Trans-Pecos volcanic field, 7, 25, 31, 211

Trinity Group, 42

Twin Peaks, 92, 93, 144, 151, 153

Valentine, 36, 122, 172, 173, 186, 192, 197, 201

Valentine basin, 122, 197, 201

Van Horn, 18, 105, 191

Van Horn Mountains, 12, 190, 192, 193, 195, 196, 199

Van Horn Mountains caldera, 192, 196, 197, 199

Van Horn Sandstone, 191

Victorio Peak Limestone, 182, 183

Village Farms, 164, 166

Washita, 42, 44, 45

Weston intrusion, 139, 144

White Mountain, 128, 129, 131

Whitetail Mountain, 113, 117, 118, 121

Whiting, Lieutenant William H.C., 74, 193, 196, 212

Wild Cherry tuffs, 26, 37, 76, 85, 86, 91, 96, 113, 118, 123, 127, 129, 131

Wild Rose Pass, 52, 62, 63, 64, 67, 105, 150

Woulfter Peak, 109

Wylie Mountains, 182, 183, 184, 188, 190, 192, 193

Y-6 Hills, 171, 177, 197